T0222388

PHYSICS
OLYMPIAD
Basic to Advanced Exercises

PHYSICS
OLYMPIAD

Basic to Advanced Exercises

The Committee of
Japan Physics Olympiad

World Scientific

NEW JERSEY · LONDON · SINGAPORE · BEIJING · SHANGHAI · HONG KONG · TAIPEI · CHENNAI

Published by

World Scientific Publishing Co. Pte. Ltd.

5 Toh Tuck Link, Singapore 596224

USA office: 27 Warren Street, Suite 401-402, Hackensack, NJ 07601

UK office: 57 Shelton Street, Covent Garden, London WC2H 9HE

Library of Congress Cataloging-in-Publication Data
Committee of Japan Physics Olympiad.
    Physics Olympiad : basic to advanced exercises / The Committee of Japan Physics Olympiad.
       pages cm
    Includes index.
    ISBN-13: 978-9814556675 (pbk. : alk. paper)
    ISBN-10: 981455667X (pbk. : alk. paper)
    1. Physics--Problems, exercises, etc. 2. Physics--Competitions. I. Title.
    QC32.C623 2013
    530.076--dc23
                                2013037572

British Library Cataloguing-in-Publication Data
A catalogue record for this book is available from the British Library.

In-house Editor: Song Yu

Typeset by Stallion Press
Email: enquiries@stallionpress.com

Printed in Singapore

Preface to the English Edition

The Committee of Japan Physics Olympiad (JPhO), a non-profit organization approved and supported by the Japanese government, has organized Physics Challenge, a domestic competition in physics, for high-school students, every year since 2005 and has also selected and sent the best five students to represent Japan in the International Physics Olympiad (IPhO) every year since 2006. The main aim of the activity of our Committee is to promote and stimulate high-school–level physics education in Japan so as to achieve a world-class standard, which we have experienced during the IPhO.

Physics Challenge consists of three stages: the First Challenge, the Second Challenge, and the Challenge Final. The First Challenge selects about 100 students from all applicants (1000~1500 in total every year); every applicant is required to take a theoretical examination (90 min, multiple-choice questions) held at more than 70 places on a Sunday in June, and to submit a report on an experiment done by himself. The subject of the experiment is announced several months before the submission deadline.

The Second Challenge is a four-day camp held in August; all students in the Second Challenge lodge together for the whole four days. Each student takes a theoretical examination and an experimental examination; both are five hours long just like the examinations in the IPhO.

The best 10–15 students who show excellent scores in the Second Challenge are nominated as candidates for the Japan team for the IPhO. They are then required to participate in a four-day winter camp at the end of December and a four-day spring camp at the end of March. They are also required to have monthly training via email; the training consists of a series of questions and takes place from September to March. At the end of the spring camp, these candidates take the Challenge Final, which consists of theoretical and experimental examinations. The best five students are then selected to form the Japan team for the IPhO.

This book contains some of the questions in the theoretical and experimental examinations of previous Physics Challenges. Elementary Problems in this book are taken from the First Challenge competitions and Advanced Problems are mostly from the Second Challenge competitions. Through these questions, we hope that high-school students would become excited and interested in modern physics. The questions from the Second Challenge reflect the process of development of physics; they ranges from very fundamental physics of junior-high-school level to the forefront of advanced physics and technology. These problems are, we believe, effective in testing the students' ability to think logically, their stamina to concentrate for long hours, their spirit to keep trying when solving intricate problems, and their interest to do science. We do not require students to learn physics by a piecemeal approach. In fact, many of the basic knowledge of physics for solving the problems are given in the questions. But, of course, since the competitions at the IPhO require fundamental knowledge and skills in physics, this book is organized in such a way that the basics are explained concisely together with some typical basic questions to consolidate the knowledge.

This book is not only meant for training students for physics competitions but also for making students excited to learn physics. We often observed that the content of physics education in high school is limited to basic concepts and it bears little relation to modern and cutting-edge science and technology. This situation may make physics class dull. Instead, we should place more emphasis on the diversity and vastness of the application of physics principles in science and technology, which is evident in everyday life as well useful for gaining a deeper understanding of our past. Therefore, we try in this book to bridge the gap between the basics and the forefront of science and technology. We hope that this book will be used in physics classes in high schools as well as in extracurricular activities.

We deeply appreciate the following people for their contributions to translating the original Japanese version into English and editing the manuscript: Kazuo Kitahara, Tadao Sugiyama, Shuji Hasegawa, Kyoji Nishikawa, Masao Ninomiya, John C. Gold Stein, Isao Harada, Akira Hatano, Toshio Ito, Kiyoshi Kawamura, Hiroshi Kezuka, Yasuhiro Kondo, Kunioki Mima, Kaoru Mitsuoka, Yusuke Morita, Masashi Mukaida, Yuto Murashita, Daiki Nishiguchi, Takashi Nozoe, Fumiko Okiharu, Heiji Sanuki, Toru Suzuki, Satoru Takakura, Tadayoshi Tanaka, Yoshiki Tanaka, and Hiroshi Tsunemi.

January 2013
The Committee of Japan Physics Olympiad

Contents

Part II. Experiment   271

Chapter 7. How to Measure and Analyze Data   273

Chapter 8. Practical Exercises   299

Appendix. Mathematical Physics   321

PART I

Theory

Chapter 1

General Physics

Elementary Problems

Problem 1.1.   The SI and the cgs systems

The units of fundamental physical quantities, such as length, mass and time, are called the **fundamental units**, from which the units of other physical quantities are derived.

In the International System of Units (SI), the unit of length is the meter (m), that of mass is the kilogram (kg) and that of time is the second (s). Other units can be composed of these fundamental units. For example, the unit of mass density is kg/m^3, because the unit of volume is m × m × m = m^3.

On the other hand, there are units composed of the gram (g), the unit of mass; the centimeter (cm), the unit of length; and the second (s), the unit of time. This system of units is called the **cgs system of units**. In the cgs system, the unit of volume is cm^3 and the unit of mass density is g/cm^3.

The unit size in the SI is not the same as that in the cgs system. For example, 1 m^3 in the SI unit is 10^6 cm^3 in the cgs unit.

How many times larger is the unit size in the SI as compared with the unit size in the cgs system for each of the following physical quantities?

Enter the appropriate numbers in the blanks below.

the unit of volume: 10^a times $a = \boxed{6}$

(1) the unit of speed: 10^i times $i = \boxed{}$

(2) the unit of acceleration: 10^j times $j = \boxed{}$

(3) the unit of force: 10^k times $k = \boxed{}$

(4) the unit of energy: 10^l times $l = \boxed{}$

(5) the unit of pressure: 10^m times $m = \boxed{}$

(the 1st Challenge)

Answer $i = 2$, $j = 2$, $k = 5$, $l = 7$, $m = 1$

Solution

(1) The unit of speed in the SI is m/s. The unit size of speed in the SI is $1\,\text{m/s} = 1 \times 10^2\,\text{cm/s}$ (because $1\,\text{m} = 1 \times 10^2\,\text{cm}$). Therefore, it is $\underline{10^2}$ times the unit size of speed in the cgs system.

(2) The unit of the acceleration in the SI is m/s². $1\,\text{m/s}^2 = 1 \times 10^2\,\text{cm/s}^2$. Therefore, the answer is $\underline{10^2}$ times.

(3) Force is "(mass) × (acceleration)", therefore the unit of force in the SI is $\text{N} = \text{kg·m/s}^2$. $1\,\text{N} = 1 \times 10^3\,\text{g} \times 10^2\,\text{cm/s}^2 = 10^5\,\text{g·cm/s}^2 = 10^5\,\text{dyn}$ (because $1\,\text{kg} = 10^3\,\text{g}$). Therefore, the answer is $\underline{10^5}$ times.

(4) Energy is "(force) × (distance)", therefore the unit of energy in the SI is $\text{J} = \text{N·m}$. $1\,\text{J} = 10^5\,\text{dyn} \times 10^2\,\text{cm} = 10^7\,\text{erg}$. Therefore, the answer is $\underline{10^7}$ times.

(5) Pressure is "(force)/(area)", therefore the unit of pressure in the SI is $\text{Pa} = \text{N/m}^2$. $1\,\text{Pa} = 10^5\,\text{dyn}/10^4\,\text{cm}^2 = 10\,\text{dyn/cm}^2$. Therefore, the answer is $\underline{10}$ times.   ∎

Problem 1.2.   The pressure due to high heels and elephants

Suppose the total weight of a person who wears high heels is $50\,\text{kg}$ and is carried only on the ends of both heels equally (assume the cross section at the end of one heel to be $5\,\text{cm}^2$). Also, suppose the total weight of an elephant is $4000\,\text{kg}$ and is carried equally on the four soles (assume the cross section of one sole to be $0.2\,\text{m}^2$). How many times larger is the pressure exerted on one sole of the elephant compared with the pressure exerted on the end of one heel of the high heels?

Choose the best answer from (a) through (f).

(a) $\frac{1}{20}$ times (b) $\frac{1}{10}$ times (c) $\frac{1}{5}$ times

(d) 5 times (e) 10 times (f) 20 times

(the 1st Challenge)

Answer (e)

Solution

It is important to express the units of physical quantities in the SI. Let the gravitational acceleration be g. The person's weight, $50\,g$, is carried on the ends of both heels equally. Hence, the pressure exerted on the end of one heel is $p_H = \frac{50g}{2\times5\times10^{-4}} = 5 \times 10^4 g\,$Pa (because $5\,\mathrm{cm}^2 = 5 \times 10^{-4}\,\mathrm{m}^2$); the pressure exerted on one sole of the elephant is $p_E = \frac{4000g}{4\times0.2} = 5 \times 10^3 g\,$Pa. Hence, the answer is $\frac{p_H}{p_E} = 10$(times). ∎

Problem 1.3. The part of the iceberg above the sea

As shown in Fig. 1.1, an iceberg is floating in the sea. Find the ratio of the volume of the part of the iceberg above the sea to the whole volume of the iceberg. Here, the density of seawater is $1024\,\mathrm{kg/m}^3$ and the density of ice is $917\,\mathrm{kg/m}^3$.

Choose the best answer from (a) through (f).

(a) 89.6% (b) 88.3% (c) 52.8% (d) 47.2% (e) 11.7% (f) 10.4%

(the 1st Challenge)

Answer (f)

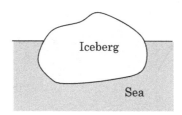

Fig. 1.1.

Solution

The buoyant force exerted on the iceberg is equal to the weight of the seawater displaced by the iceberg.

Let the whole volume of the iceberg be V, the volume of the seawater displaced by the iceberg be v, the density of seawater be $\rho_s = 1024\,\mathrm{kg/m}^3$, the density of ice be $\rho_i = 917\,\mathrm{kg/m}^3$ and the gravitational acceleration be g. Since the forces on the iceberg are balanced, $\rho_i V g = \rho_s v g$.

Hence, the ratio of the volume of the part above the sea to the whole volume of the iceberg is $\frac{V-v}{V} = 1 - \frac{v}{V} = 1 - \frac{\rho_i}{\rho_s} = 1 - \frac{917}{1024} = 0.104$, i.e., 10.4%.   ∎

Supplement
The buoyancy on a body equals the resultant force due to the pressure exerted by the surrounding fluid

The pressure on a body of volume V due to its surrounding fluid (whose density is ρ) acts perpendicularly to the boundary surface between the body and the fluid (see Fig. 1.2(a)).

Since the fluid pressure at a deep location is greater than that at a shallow location, the resultant force due to the pressure on the boundary surface points upward. This resultant force is the buoyancy, denoted as F, acting on the body.

Let us consider a region of fluid with the same volume V as the body (see Fig. 1.2(b)). The buoyancy, F, acting on this region is equal to the force exerted vertically on the body by its surrounding fluid.

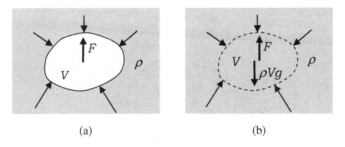

(a)   (b)

Fig. 1.2.

Simultaneously, a gravitational force of $\rho V g$ acts on this region of volume V. Therefore, it turns out that the magnitude of the buoyancy is given by $F = \rho V g$ due to the balance of the forces acting on the region of the fluid of volume V.

For a body floating in a fluid, the magnitude of the buoyancy acting on the body is equal to the magnitude of the gravitational force on the fluid displaced by the part of the body submerged in the fluid.

Problem 1.4. The altitude angle of the Sun

Suppose the length of the meridian from the North Pole to the Equator is 10000 km. What is the difference between the altitude angle of the Sun at Amagi-san in Izu and that in Niigata City, which lies 334 km north of Amagi-san when the Sun crosses the meridian that passes through both?

Choose the best answer from (a) through (f).

    (a) $1°$ (b) $1.5°$ (c) $3°$ (d) $4.5°$ (e) $6°$ (f) $12°$

                                                  (the 1st Challenge)

Answer (c)

Hint At the instant when the Sun crosses the meridian, the difference between the altitude angles of the Sun is equal to the difference between the latitudes of the two locations.

Solution

Let angle θ be the difference between the altitude angle at Amagi-san and that at Niigata City. Let point A be Amagi-san, point N be Niigata City and point O be the center of the Earth. We further define the angle $\angle AON = \theta$ (see Fig. 1.3).

The altitude angles of the Sun at points N and A at the instant when the Sun crosses the meridian are equal to the angles between the southern tangents to the Earth and the lines pointing toward the Sun N→S and A→S, respectively (see Fig. 1.3). Hence, the difference between the altitude angles at points A and N is $\phi_A - \phi_N = \theta$, where

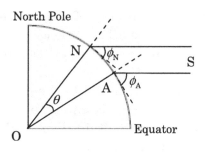

Fig. 1.3.

ϕ_A and ϕ_N are the altitude angles of the Sun at points A and N, respectively.

Given that the meridian length from the North Pole to the Equator is 10000 km and the distance between Amagi-san and Niigata City is 334 km,

$$\theta = \frac{334}{10000} \times 90° = \underline{3.0°}.$$ ■

Advanced Problems

Problem 1.5.   Dimensional analysis and scale transformation

I. Once the fundamental units, namely, the standard units in length, mass and time, are specified, the units of any other physical quantities can be determined in terms of (combinations of) the fundamental units. Such a combination is called the **dimension** of the physical quantity of concern. In an equation that represents a relation between two physical quantities, the dimensions on both sides of the equation must be the same. By investigating dimensions, it is possible to examine the relation between a physical quantity and other physical quantities, except for some (dimensionless) numerical factor. This investigation is called **dimensional analysis**.

We represent the dimension of mass by [M], the dimension of length by [L] and the dimension of time by [T]. Then, we can study some physical phenomena in terms of these dimensions.

Suppose the speed of sound, v_s, is expressed as

$$v_s = k\,p^a\rho^b, \tag{1.1}$$

(k is a dimensionless coefficient; a and b are some numbers.).

in terms of atmospheric pressure, p, and density of air, ρ. Here, the dimensions of v_s, p and ρ are $[v_s] = [\mathrm{LT}^{-1}]$, $[p] = [(\mathrm{MLT}^{-2})/\mathrm{L}^2] = [\mathrm{ML}^{-1}\mathrm{T}^{-2}]$ and $[\rho] = [\mathrm{M}/\mathrm{L}^3] = [\mathrm{ML}^{-3}]$, respectively. Such that the equality of the dimensions on both sides of Eq. (1.1) is

$$[\mathrm{LT}^{-1}] = [\mathrm{ML}^{-1}\mathrm{T}^{-2}]^a[\mathrm{ML}^{-3}]^b = [\mathrm{M}^{a+b}\mathrm{L}^{-a-3b}\mathrm{T}^{-2a}].$$

Therefore,

$$0 = a + b, \quad 1 = -a - 3b, \quad -1 = -2a,$$

$$\therefore\ a = \frac{1}{2} \quad \text{and} \quad b = -\frac{1}{2}$$

(1) Let's consider the lift of an airplane.

We may model the wing of an airplane by a rectangular plane of length W and width L. Suppose this airplane flies in the atmosphere at a speed, v, relative to the atmosphere whose density is ρ.

Since the lift on an airplane, F, is proportional to the length of its wing, we may write

$$\frac{F}{W} = k\rho^a v^b L^c$$

(k is a dimensionless coefficient; a, b and c are some numbers).

Find the values of indices a, b and c by dimensional analysis.

(2) The airplane takes off with a speed of 250 km per hour, and flies at a speed of 900 km per hour at an altitude of 10000 m. Suppose the lift acting on the airplane at this altitude is equal to that at the moment when the airplane takes off from the ground. Then, estimate the ratio $\frac{\rho_1}{\rho_0}$ where ρ_1 is the density of the air at the altitude of 10000 m and ρ_0 is the density on the ground.

II If a physical quantity is expressed in terms of powers of other physical quantities, we can study physical laws under a scale transformation. Now, suppose the scale of length is transformed as $r \to r_1 = \alpha r$ and the scale of time as $t \to t_1 = \beta t$ where α and β are some numbers.

(3) Then, velocity, V, and acceleration, A, are transformed as

$$V \to V_1 = \alpha^i \beta^j V \quad \text{and} \quad A \to A_1 = \alpha^k \beta^l A,$$

respectively. Find the numerical values of i, j, k and l.

(4) Let a miniature representation that has a scale factor of $\frac{1}{100}$ in length fall and record the state of affairs on a video. If the acceleration is unchanged by the scale transformation, the state of affairs looks real.

How fast should the playback speed of the videotape be as compared with the original speed if we want the fall of the miniature representation to look like that of the real object?

(the 2nd Challenge)

Solution

(1) $[F] = [\mathrm{MLT}^{-2}], [W] = [\mathrm{L}], [\rho] = [\mathrm{ML}^{-3}], [v] = [\mathrm{LT}^{-1}], [L] = [\mathrm{L}]$
Then

$$\frac{[\mathrm{MLT}^{-2}]}{[\mathrm{L}]} = [\mathrm{ML}^{-3}]^a [\mathrm{LT}^{-1}]^b [\mathrm{L}]^c,$$

$$[\mathrm{MT}^{-2}] = [\mathrm{M}^a \mathrm{L}^{-3a+b+c} \mathrm{T}^{-b}].$$

Therefore,

$$1 = a, \quad 0 = -3a + b + c, \quad -2 = -b,$$
$$\therefore \ a = \underline{1}, \quad b = \underline{2}, \quad c = \underline{1}.$$

(2) From the result of part (1),

$$\frac{F}{W} = k\rho^1 v^2 L^1.$$

Since the lift acting on the airplane at the altitude of $10000\,\mathrm{m}$ equal that at the moment when the airplane takes off from the ground, $\rho \propto v^{-2}$

$$\therefore \ \frac{\rho_1}{\rho_0} = \frac{900^{-2}}{250^{-2}} = \left(\frac{250}{900}\right)^2 \approx \underline{0.077}.$$

In this question, the number of significant digits is not mentioned. In such a case, you should put your answer in 2 or 3 significant digits.

(3) Velocity is the rate of change of the position of an object, and acceleration is the rate of change of the velocity of an object. Therefore, V and A are transformed as $V \to V_1 = \alpha\beta^{-1}V$ and $A \to A_1 = \alpha\beta^{-2}A$, respectively.

$$\therefore\ i = \underline{1}, \quad j = \underline{-1}, \quad k = \underline{1}, \quad l = \underline{-2}.$$

(4) Since the acceleration of fall is unchanged by the scale transformation, $\alpha\beta^{-2} = 1$. Since $\alpha = \frac{1}{100}$, we have $\beta = \alpha^{\frac{1}{2}} = (\frac{1}{100})^{\frac{1}{2}} = \frac{1}{10}$. In other words, the playback speed of the video should be $\frac{1}{10}$ times the original speed. ∎

Problem 1.6. Why don't clouds fall?

A cloud is a collection of water droplets that float in the atmosphere. The diameters of the water droplets are about $3\,\mu m$ to $10\,\mu m$ ($1\,\mu m = 1\times10^{-6}$ m). These water droplets are very small; their density is equal to that of water but is much larger than that of the atmosphere. Hence, it is a mystery how clouds float in the atmosphere.

Why don't clouds fall? Also, how do water droplets in a cloud fall as rain? Answer the following questions:

(1) Suppose a mass of air containing plenty of water vapor was made in the atmosphere. Describe in about 80 words the process by which this mass of air becomes a cloud in the sky.
(2) Describe in about 50 words why a cloud does not fall.
(3) Describe the process of the formation of rain in a cloud by considering the relative motion between water droplets and the air containing plenty of water vapor in the cloud.

(the 2nd Challenge)

Solution

(1) Because water vapor is less dense than the surrounding air, the mass of the air containing plenty of water vapor rises

upwards. As the mass of the air rises upwards, it expands adiabatically (because pressure decreases as altitude increases) and the temperature of the mass decreases. When the vapor pressure exceeds the saturated vapor pressure, which decreases as temperature decreases, a part of the water vapor condenses and forms minute drops of water. Thus, a cloud is formed.

(2) A cloud is made up of minute drops of water and water vapor. Water vapor is less dense than air. However, after averaging the densities of water droplets and water vapor, the density of a cloud is equal to that of the air. As a result, a cloud does not fall.

(3) In a cloud, the dense droplets of water descend and the less-dense water vapor rises. In this relative motion, viscosity plays an important role. The viscous force acting on droplets of water is proportional to the product of the radius of the droplet and its speed relative to the surrounding air (This is called **Stokes' law**). It acts in the direction opposite to the velocity of the droplet. In comparison, the weight of each droplet of water is proportional to the cube of its radius. Hence, when the droplets of water are small, their speeds relative to the water vapor is slow and the droplets of water stay in the cloud; when the droplets of water become large, their falling speeds become fast, and the droplets of water rush out of the cloud and fall down as rain. ∎

Supplement

In writing the answer above, we focus on the following:

- The air in a cloud is filled with minute drops of water and saturated water vapor.
- Water vapor is less dense than air because the molecular weight of a water molecule (H_2O) is 18 and is smaller than the aerial "average molecular weight" of 29.
- When a volume of gas rises upwards, the surrounding atmospheric pressure decreases. Since air hardly conducts heat, the ascending gas adiabatically expands, and consequently, the temperature of the gas decreases.

- The saturated pressure of water vapor decreases as temperature decreases. When the vapor pressure in the cloud exceeds the saturated value, the vapor changes into liquid water.

Moreover, a detailed description of the process of enlargement of the water droplets are as follows:

Because water droplets in the atmosphere easily acquire electrical charges of the same sign, they are repelled from one another. As a result, they do not combine with one another and become too large. However, when the water droplets discharge their electrical charges via thunderbolts, the repulsive forces between them disappear. They can, then, combine and rapidly become large. And then, they fall as rain drops. This is what happens in a thunderstorm.

A water droplet may absorb its surrounding water vapor and grow larger. The rate at which water in the droplet vaporizes is large when the water droplet is small, and it is possible that the water droplet becomes smaller and disappears. However, once the radius of the water droplet becomes larger than a certain critical value, the water droplet grows rapidly as water vapor condenses on its surface.

Chapter 2

Mechanics

Elementary Course

2.1. Motion with a Constant Acceleration

The rate of change of the displacement of a body with respect to time is called the **velocity** of the body and the rate at which the velocity of the body changes with respect to time is called the **acceleration** of the body.

Suppose a body moves along the x-axis with a constant acceleration, a. If the body has a velocity of $\nu = v_0$ at point $x = x_0$ at time $t = 0$, the velocity, v, at time t is

$$v = v_0 + at. \tag{2.1}$$

Figure 2.1 shows v as a function of t. From the fact that the displacement travelled, Δx, during time interval Δt is $v\Delta t$, it is deduced that the displacement, x, at time t is given by the area of the shadowed trapezoid in Fig. 2.1. Thus, we find

$$x = x_0 + \frac{1}{2}[v_0 + (v_0 + at)] \times t$$

$$= x_0 + v_0 t + \frac{1}{2}at^2. \tag{2.2}$$

After eliminating t from Eqs. (2.1) and (2.2), we obtain

$$v^2 - v_0^2 = 2a(x - x_0). \tag{2.3}$$

Example 2.1. Velocity, v, is defined as $v \equiv \frac{dx}{dt}$, a derivative of x with respect to t, and acceleration, a, is defined as $a \equiv \frac{dv}{dt}$. By integrating these two expressions, derive Eqs. (2.1) and (2.2) for the case of constant acceleration.

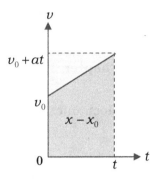

Fig. 2.1.

Solution

When a is constant, integrating $a = \frac{dv}{dt}$ with respect to t yields

$$v = \int a\,dt = at + C_1, \quad (C_1 \text{ is an integral constant}).$$

By using the initial condition $v = v_0$ at $t = 0$, we have $C_1 = v_0$. Thus, we obtain Eq. (2.1).

Integrating $v = \frac{dx}{dt}$ with respect to t yields

$$x = \int v\,dt = \int (v_0 + at)dt = v_0 t + \frac{1}{2}at^2 + C_2,$$

$$(C_2 \text{ is an integral constant}).$$

By using the initial condition $x = x_0$ at $t = 0$, we have $C_2 = x_0$. Thus, we obtain Eq. (2.2). ∎

2.1.1. *Projectile Motion*

When air resistance is negligible, a body moves with the constant acceleration due to gravity, g, which points in the downward direction. In addition, it moves at a constant speed in the horizontal direction, since no force acts on it horizontally. Thus, the body moves in a parabolic path.

As shown in Fig. 2.2, we take the origin O to be a point on the ground, x to be the displacement in the horizontal direction and y to be the displacement in the vertical direction. Suppose a body moving

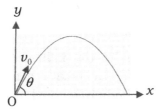

Fig. 2.2.

at speed v_0 is launched at an angle, θ, to the x-axis at $t = 0$. The coordinates (x, y) of the body at time t are

$$x = v_0 \cos \theta \cdot t, \quad y = v_0 \sin \theta \cdot t - \frac{1}{2} g t^2.$$

After eliminating t from these equations, we have

$$y = x \tan \theta - \frac{g}{2 v_0^2 \cos^2 \theta} x^2.$$

This equation implies that the body moves in a parabolic path.

2.2. Equation of Motion

When a force, f, acts on a body, the body has an acceleration, a, proportional to the force. When we use $\frac{1}{m}$ as the coefficient of proportionality, we have (see Fig. 2.3)

$$a = \frac{1}{m} f \iff ma = f. \tag{2.4}$$

Equation (2.4) is called the **equation of motion**. The equation of motion is not derived from any other law: it is one of the fundamental laws in Newtonian mechanics.

Fig. 2.3.

2.3. The Law of Conservation of Energy

2.3.1. *Work and Kinetic Energy*

Suppose a body moves under the influence of a constant force (Fig. 2.4). In vector notation, a force is denoted as $\vec{f}$. When the displacement vector of the body is denoted as $\vec{r}$, we define the work done by the force, $\vec{f}$, on the body as

$$W \overset{\text{def}}{=} \vec{f} \cdot \vec{r} = f\, r \cos \theta. \tag{2.5}$$

The product at the center of this equation is called the inner product of $\vec{f}$ and $\vec{r}$ and its value is given by the rightmost expression, where f and r are magnitudes of the vectors $\vec{f}$ and $\vec{r}$, respectively, and θ is the angle between them.

We consider a body of mass m moving along the x-axis under the influence of a constant force, f, and passing a point, x_1, with a velocity, v_1, and then passing another point, x_2, with another velocity, v_2, as shown in Fig. 2.5. In terms of acceleration, a, the equation of motion is written as $ma = f$. Here, a is constant, because f is a constant force. By setting $x_0 = x_1$, $x = x_2$, $v_0 = v_1$ and $v = v_2$ in Eq. (2.3), we obtain

$$v_2^2 - v_1^2 = 2a(x_2 - x_1).$$

Fig. 2.4.

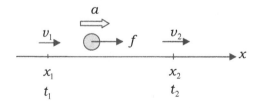

Fig. 2.5.

After multiplying both sides of this equation by $\frac{m}{2}$ and using the equation of motion $f = ma$, we have

$$\frac{1}{2}mv_2^2 - \frac{1}{2}mv_1^2 = ma(x_2 - x_1) = f(x_2 - x_1). \qquad (2.6)$$

Given a body of mass m that moves at speed v, we call the quantity K defined as

$$K \stackrel{\text{def}}{=} \frac{1}{2}mv^2, \qquad (2.7)$$

the **kinetic energy** of the body. The left side of Eq. (2.6) represents the change in the kinetic energy during the motion from x_1 to x_2 and the rightmost expression is the work done during this motion. Therefore, Eq. (2.6) is interpreted as

> **"the change in the kinetic energy of a body is equal to the work done by the force on the body."** (2.8)

Although we derive Theorem (2.8) for a one-dimensional system under the influence of a uniform force, by using vector algebra, the theorem can be deduced from the equation of motion for the three-dimensional motion of a body under the influence of a spatially varying force.

Example 2.2. Suppose a body moves along the x-axis under the influence of an x-dependent force. Derive Theorem (2.8) from the equation of motion by integrating it.

Solution

Consider, again, the one-dimensional motion shown in Fig. 2.5, but assume that the force acting on the body depends on x.

Substituting $a = \frac{dv}{dt}$ into Eq. (2.4) yields

$$m\frac{dv}{dt} = f.$$

Then, multiply both sides of this equation by $v = \frac{dx}{dt}$ and integrate it from t_1 to t_2. After converting the integral variables, we have

$$\text{the left side} = \int_{t_1}^{t_2} mv\frac{dv}{dt}dt = \int_{v_1}^{v_2} mv\,dv = \frac{1}{2}mv_2^2 - \frac{1}{2}mv_1^2,$$

$$\text{the right side} = \int_{t_1}^{t_2} f\frac{dx}{dt}dt = \int_{x_1}^{x_2} f\,dx = W(x_1 \to x_2).$$

On the right side, $f\,dx$ is the work done by f as the body travels an infinitesimal displacement of dx. Hence, $W(x_1 \to x_2)$ denotes the work done by the varying force acting on the body during the displacement from x_1 to x_2. By equating the above two expressions, we obtain

$$\frac{1}{2}mv_2^2 - \frac{1}{2}mv_1^2 = W(x_1 \to x_2). \tag{2.9}$$

Therefore, Theorem (2.8) is valid even for one-dimensional motions with varying forces. ∎

2.3.2. *Conservative Forces and Non-conservative Forces*

In general, when the work done by a force as an object travels over an arbitrary displacement depends only on the starting and ending positions of the body, that force is called a **conservative force**.

Consider the one-dimensional motion of a body along the x-axis under the influence of a constant force, f, that points to the positive direction of the x-axis (see Fig. 2.5). The work done by f as the body moves from x_1 to x_2 $(x_1 < x_2)$ is

$$W_1(x_1 \to x_2) = f(x_2 - x_1). \tag{2.10}$$

Whereas when the body moves first from x_1 to x_3 $(x_1 < x_2 < x_3)$ and then from x_3 to x_2, the work done by f is

$$W_2(x_1 \to x_2) = f(x_3 - x_1) + f(x_2 - x_3)$$
$$= f(x_2 - x_1) = W_1(x_1 \to x_2)$$

As x_3 is an arbitrary point, this relation implies that the work done by a constant force is independent of the path taken to travel from x_1 to x_2 and is determined only by the starting and ending positions of the body. Therefore, a constant force f is conservative. The gravitational and the elastic forces are two examples of conservative forces.

A force that is not conservative is called a **non-conservative force**. The work done by a non-conservative force on a body depends on the path of motion as well as the initial and final positions of the body.

An example of non-conservative force is the frictional force. The direction of a kinetic-frictional force is always opposite to the direction of motion, so the θ in Eq. (2.5) should be set to $180°$. Hence, the work done by a kinetic-frictional force of a constant magnitude, f', as an object moves from x_1 to x_2 is

$$W_1'(x_1 \to x_2) = -f'(x_2 - x_1).$$

On the other hand, suppose a solid body slides along the x-axis first from x_1 to x_3 and then from x_3 to x_2 on a solid horizontal surface with a kinetic-frictional force of a constant magnitude, f'. The first displacement is $x_3 - x_1$ and the second displacement is $x_2 - x_3$. Hence, the work done to travel the total displacement from x_1 to x_2 is

$$W_2'(x_1 \to x_2) = -f'(x_3 - x_1) - f'(x_3 - x_2)$$
$$= -f'(2x_3 - x_1 - x_2) \neq W_1'(x_1 \to x_2),$$

where we have used the fact that the work done by a kinetic-frictional force is negative. This result implies that the work done by a kinetic-frictional force depends on the path travelled: a kinetic-frictional force is non-conservative.

2.3.3. *Potential Energy*

When a body moves from point P to point O under the action of a conservative force, $\vec{f}$, the work done by $\vec{f}$, $W(P \to O)$, is determined by both the positions of P and O. We define the **potential energy** possessed by the body at P, $U(\mathrm{P})$, as

$$U(\mathrm{P}) \equiv W(\mathrm{P} \to \mathrm{O}). \qquad (2.11)$$

Note that the potential energy at P, $U(\mathrm{P})$, depends on that at O, which is called a **reference point**. Therefore, an arbitrary constant, C, dependent on the choice of the reference point can be added to the potential energy.

We cannot define a potential energy for any non-conservative forces.

2.3.4. *Examples of Potential Energy*

Gravitational Potential Energy

We usually choose the ground as the reference point for defining the **gravitational potential energy** of a body so that its value at a height, h_0, is equal to the work done by the gravity as the body moves down from that height to the ground, as shown in Fig. 2.6. The gravitational potential energy of a body is calculated in terms of the **gravitational acceleration**, g:

$$U(h_0) = \int_{h_0}^{0} (-mg)dy = mg \int_{0}^{h_0} dy = mgh_0. \qquad (2.12)$$

Elastic Potential Energy

As shown in Fig. 2.7, a light spring with a spring constant of k is fixed at the left end and a particle of mass m is attached to the right end. They are constrained to move only along the x-axis. We take the reference point for defining the potential energy of the particle

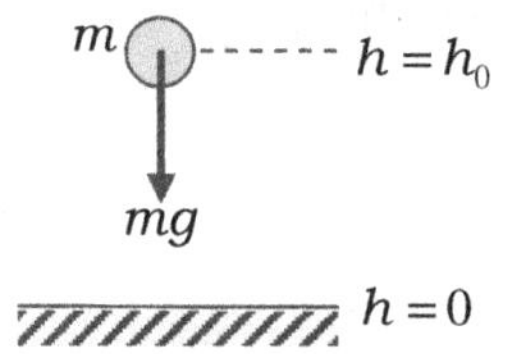

Fig. 2.6.

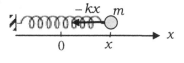

Fig. 2.7.

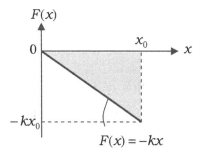

Fig. 2.8.

to be the position where the spring does not exert an elastic force on the particle. We choose this reference point to be the origin of the x-axis and the x-axis to point to the right along the length of the spring. The elastic force at a position, x, is $F(x) = -kx$. Then, the elastic potential energy of the particle at $x = x_0$, $U(x_0)$, is equal to the work done by the elastic force of the spring as the particle moves from $x = x_0$ to $x = 0$, $W(x_0 \to 0)$. The value of $W(x_0 \to 0)$ is given by the area of the shadowed triangle in Fig. 2.8. Thus, we obtain (see Example 2.3)

$$U(x_0) \stackrel{\text{def}}{=} W(x_0 \to 0) = \frac{1}{2}kx_0^2. \tag{2.13}$$

So far, we employed the viewpoint that the elastic potential energy is possessed by the particle connected to the spring. Instead, it is possible to consider that the energy is stored in the deformed spring. Its energy is called the elastic energy of the spring.

Example 2.3. Derive Eq. (2.13) by integration.

Solution

From the equation $F(x) = -kx$, we have

$$U(x_0) = W(x_0 \to 0) = \int_{x_0}^{0} (-kx)dx = \int_{0}^{x_0} kx\,dx = \frac{1}{2}kx_0^2.$$

∎

2.3.5. *The Law of Conservation of Mechanical Energy*

As illustrated in Fig. 2.9 with a reference point, x_0, consider a body of mass m moves along the x-axis under the action of a conservative force, f. We denote the velocity at $P_1(x = x_1)$ as v_1 and the velocity at $P_2(x = x_2)$ as v_2. Then, Eq. (2.9) can be written as

$$\frac{1}{2}mv_2^2 - \frac{1}{2}mv_1^2 = W(x_1 \to x_2) = W(x_1 \to x_2 \to x_0) - W(x_2 \to x_0)$$
$$= U(x_1) - U(x_2).$$

Hence, we have

$$\frac{1}{2}mv_2^2 + U(x_2) = \frac{1}{2}mv_1^2 + U(x_1). \tag{2.14}$$

This shows that the sum of the kinetic energy and the potential energy at $x = x_2$ is equal to that at $x = x_1$. The sum of these two energies is called the **mechanical energy**, and the equality in Eq. (2.14) is called the **law of conservation of mechanical energy**. So far, we only considered work and energies associated with one-dimensional motions, but these quantities can also be defined

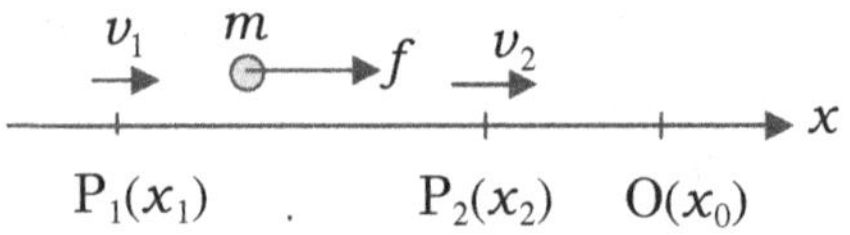

Fig. 2.9.

for three-dimensional motions. Consequently, various dynamical processes in three-dimensional space can be analyzed using the law of conservation of energy.

2.3.6. *Energy Transfer between Interacting Bodies*

The energy of a body is a measure of the ability of the body to do work to other bodies. The **kinetic energy** of a body is the energy that the body has as a result of its motion and the **potential energy** of a body is the energy that the body has as a result of its position.

When a body exerts a force on another body and the force does work, the mechanical energy of the former body decreases by the same amount as the work done by the force, while the mechanical energy of the latter body increases also by the same amount as the work done by the force.

Example 2.4. Two bodies of masses m_1 and m_2 are joined by a string that passes over a pulley (Fig. 2.10). At $t = 0$, the body of mass m_1 passes point O rightward at a speed, v_0, whereas the body of mass m_2 passes point O' upward at the same speed. We denote the

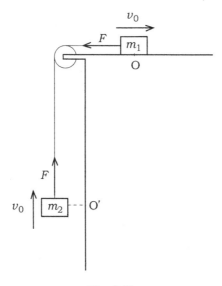

Fig. 2.10.

acceleration of the body of mass m_1 as a, which is taken as positive when directed to the right. We assume that m_2 is larger than m_1 so that a has a negative value.

(1) Write the equations of motion for the two bodies, denoting the magnitude of the tension in the string as F.

(2) Derive expressions for the coordinate and velocity of the body m_1 as a function of time, t. Give an expression for F in terms of m_1, m_2 and g.

(3) What is the time, T, when the two bodies start to move in the directions opposite to the velocities at $t = 0$? What is the distance, L, travelled by each of the two bodies at time T? Give the expressions for T and L in term of m_1, m_2, g and v_0.

(4) Show that the kinetic energy of the body of mass m_1 at $t = 0$ is converted to the work done to raise the body of mass m_2 by distance L.

(5) Show that the work in the preceding problem is converted to the mechanical energy of the body of mass m_2 after it is raised by distance L.

Solution

(1) The equation of motion for the body of mass m_1 is

$$m_1 a = -F.$$

If the acceleration of the body of mass m_2 is taken as positive when directed upward, it is equal to a. Hence, the equation of motion for the body of mass m_2 is

$$m_2 a = F - m_2 g.$$

(2) From these two equations, we have

$$F = \frac{m_1 m_2}{m_1 + m_2} g, \tag{2.15}$$

and

$$a = -\frac{m_2}{m_1 + m_2} g. \tag{2.16}$$

We denote the velocity of the body of mass m_1 at time t as $v_1(t)$ and its displacement at time t as $x_1(t)$. After substituting Eq. (2.16) into Eqs. (2.1) and (2.2) and using the initial conditions $v_1(0) = v_0$ and $x_1(0) = 0$, we have

$$v_1 = v_0 - \frac{m_2}{m_1 + m_2} gt, \tag{2.17}$$

$$x_1 = v_0 t - \frac{1}{2} \frac{m_2}{m_1 + m_2} gt^2. \tag{2.18}$$

(3) Substituting $v_1 = 0$ at $t = T$ into Eq. (2.17) yields

$$T = \frac{m_1 + m_2}{m_2} \frac{v_0}{g}. \tag{2.19}$$

Substituting the above expression for T into t on the right side of Eq. (2.18) yields

$$L \stackrel{\text{def}}{=} x_1(T) = \frac{1}{2} \frac{m_1 + m_2}{m_2} \frac{v_0^2}{g}. \tag{2.20}$$

(4) The body of mass m_1 has a kinetic energy of $\frac{1}{2} m_1 v_0^2$ at $t = 0$ and loses this amount of kinetic energy during the time interval T. At the same time, the work done by the tension in the string, F, on the body of mass m_2 is FL. From Eqs. (2.15) and (2.20), this work is evaluated as

$$FL = \frac{m_1 m_2}{m_1 + m_2} g \cdot \frac{1}{2} \frac{m_1 + m_2}{m_2} \frac{v_0^2}{g} = \frac{1}{2} m_1 v_0^2.$$

This is exactly equal to the kinetic energy of the body of mass m_1 at $t = 0$. This fact can be interpreted as follows: the body of mass m_1 has the ability to do work on the body of mass m_2 through the string.

(5) If we choose the position of the body of mass m_2 at $t = 0$ as the reference point for defining the potential energy, its potential energy at $t = T$ is $m_2 g L = \frac{1}{2}(m_1 + m_2)v_0^2$. Since this body

has a kinetic energy of $\frac{1}{2}m_2v_0^2$ at $t = 0$, the increment in the mechanical energy of this body after it is raised by distance L is $\frac{1}{2}m_1v_0^2$, which is equal to the work done by tension F during this motion. ∎

2.3.7. *Work Done by Non-conservative Forces*

In Fig. 2.11, suppose both a conservative force, f, and a non-conservative force, f', act on a body while the body moves from point P$_1$ to point P$_2$. Then, we cannot define any potential energy for f', and hence, the work done by f', W', is not contained in the mechanical energy of the body. As a result, we can write

$$\frac{1}{2}mv_2^2 - \frac{1}{2}mv_1^2 = W(x_1 \to x_2) + W' = U(x_1) - U(x_2) + W',$$

$$\therefore \left(\frac{1}{2}mv_2^2 + U(x_2)\right) - \left(\frac{1}{2}mv_1^2 + U(x_1)\right) = W'. \tag{2.21}$$

This equation shows that

"the change in the mechanical energy of a body
is equal to the work done by the non-conservative
forces acting on the body" (2.22)

In other words, when a kinetic-frictional force acts on a body, the mechanical energy of the body decreases by an amount equal to the absolute value of the work done by the friction force.

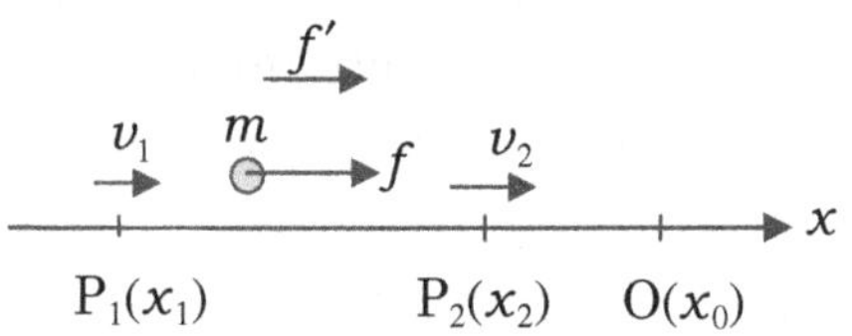

Fig. 2.11.

2.4. Newton's Law of Universal Gravitation and Kepler's Laws

When we consider the motion of stars, we should apply the law of universal gravitation. This law is derived from Kepler's laws, which are results of observation and are considered one of the fundamental laws that we should keep in mind when discussing mechanical problems.

2.4.1. *Newton's Law of Universal Gravitation*

The gravitational force acting between two stars of masses M and m separated by a distance, r, as shown in Fig. 2.12, is

$$F = -G\frac{Mm}{r^2}. \tag{2.23}$$

The negative sign of the force implies that the force is attractive. Here, the constant $G = 6.67 \times 10^{-11}$ N·m^2/kg^2 is called the **universal gravitational constant**. Equation (2.23) is **Newton's law of universal gravitation**.

Fig. 2.12.

The law of universal gravitation is applied to a system of two point masses. (Point mass is an ideal body in which its mass is considered concentrated at its center.) The gravitational force between two bodies with spherically symmetric mass distributions, however, can be computed as though both bodies are point masses. Hence, we use Eq. (2.23) when we calculate the gravitational force between two stars, because the mass distributions of stars are usually spherically symmetric.

2.4.2. *Gravitational Potential Energy*

Suppose star A of mass m is at a distance, r_0, from star B of mass M. When the reference point for defining of the gravitational potential

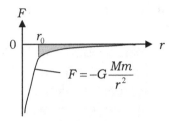

Fig. 2.13.

energy of star A is taken to be a point at infinity, its gravitational potential energy is equal to the work done by the gravitational force when star A moves from its location (that is at distance r_0 from star B) to infinity, $W(r_0 \to \infty)$. Its gravitational potential energy at position r_0, $U(r_0)$, is given by the negative of the shadowed area surrounded by the curve $F = -G\frac{Mm}{r^2}$, the r-axis and the vertical line $r = r_0$ in Fig. 2.13. By the method of integration, we calculate the area as

$$U(r_0) = -G\frac{Mm}{r_0}. \tag{2.24}$$

Example 2.5. Derive Eq. (2.24).

Solution

After integrating the gravitational force given by Eq. (2.23) from r_0 to ∞, we have

$$U(r_0) \overset{\text{def}}{=} W(r_0 \to \infty) = \int_{r_0}^{\infty} \left(-G\frac{Mm}{r^2}\right) dr = -GMm \int_{r_0}^{\infty} \frac{dr}{r^2}$$

$$= -GMm \left[-\frac{1}{r}\right]_{r_0}^{\infty} = -\frac{GMm}{r_0}. \qquad \blacksquare$$

2.4.3. *Kepler's Law*

Kepler's law is a set of the following three empirical laws.

> **Kepler's first law**: Every planet moves in an elliptical orbit that has the Sun at one of its foci.
>
> **Kepler's second law**: A line from the Sun to a planet sweeps out equal areas during equal intervals of times (Fig. 2.14).

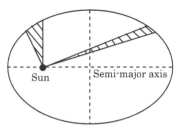

Fig. 2.14.

Kepler's third law: The square of the period of a planet is proportional to the cube of the semi-major axis of its elliptical orbit (Fig. 2.14).

Elementary Problems

Problem 2.1. A ball falling from a bicycle

Andy who is riding a bicycle drops a ball while keeping his arm at rest. Betty who is standing on the ground watches it. Choose from (a) through (e) in Fig. 2.15 the best path for the trajectory of the ball as seen by Betty.

(the 1st Challenge)

Fig. 2.15.

Answer (a)

Hint The falling ball has the same initial velocity as Andy.

Solution

Just after Andy drops a ball while keeping his arm at rest, the ball has the same velocity in the horizontal direction as the bicycle. If air resistance is negligible, the ball moves at the initial velocity horizontally in the forward direction. In the vertical direction, the ball falls freely at an initial velocity of zero with a constant gravitational acceleration of g. As a result, Betty observes that the ball falls along a parabolic trajectory in the forward direction like (a).   ■

Problem 2.2.   A ball thrown off a cliff

Suppose a ball is thrown off a cliff with the same initial speed, v_0, in the following three manners (see Fig. 2.16):

A. The ball is thrown upward.
B. The ball is thrown horizontally.
C. The ball is thrown downward.

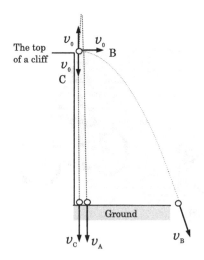

Fig. 2.16.

Here, v_A, v_B and v_C denote the speeds of the ball at the moment when it reaches the ground in cases A, B and C, respectively. Choose from the following (a) through (f) the best relation between v_A, v_B, and v_C, assuming that air resistance is negligible.

(a) $v_A > v_B > v_C$ (b) $v_C > v_B > v_A$ (c) $v_B > v_A > v_C$
(d) $v_C > v_A > v_B$ (e) $v_A = v_C > v_B$ (f) $v_A = v_B = v_C$

(the 1st Challenge)

Answer (f)

Hint Refer to the law of conservation of mechanical energy.

Solution

Let the height of the cliff from the ground be h and the gravitational acceleration be g. Let us take the horizontal axis to be the x-axis and the vertical axis to be the y-axis.

A: The ball moves in a straight line along the y-axis with a gravitational acceleration of g. The y-component of its velocity at the instant it reaches the ground is denoted by $-v_A$. By setting $v = -v_A$, $a = -g$, $x = 0$ and $x_0 = h$ in Eq. (2.3), we obtain

$$v_A = \sqrt{v_0^2 + 2gh}.$$

C: Motion in case C is different from that in case A in the sign of the initial velocity and in the notation of the final velocity. Replacing v_A with v_C as well as v_0 with $-v_0$ in the preceding equation yields

$$v_C = \sqrt{v_0^2 + 2gh}.$$

B: The velocity vector of the ball at the instant when it reaches the ground is denoted by (v_{Bx}, v_{By}). Since the horizontal motion is uniform, we have $v_{Bx} = v_0$. The vertical motion is a free fall with an initial velocity of zero and is different from the linear motion in case C in the direction of the initial velocity. Replacing v_0 and

v_C in the preceding equation with 0 and $-v_{By}$, respectively, we obtain

$$v_{By} = -\sqrt{2gh}.$$

Hence, we obtain

$$v_B = \sqrt{v_{Bx}^2 + v_{By}^2} = \sqrt{v_0^2 + 2gh}.$$

From the above, we have

$$v_A = v_B = v_C.$$

Alternative solution

In all of the three cases, the ball is initially located at the same height, h, and moving at the same speed, v_0. Using the law of conservation of energy, we get the velocity of the ball, v, at the instant it reaches the ground for each of the cases as follows:

$$\frac{1}{2}mv^2 = \frac{1}{2}mv_0^2 + mgh \quad \therefore v = \sqrt{v_0^2 + 2gh}.$$

Therefore, we obtain $v_A = v_B = v_C$. ■

Problem 2.3.　The trajectory of a ball

Suppose air resistance is negligible, consider the motion of two identical balls that are simultaneously thrown with the same initial speed toward targets 1 and 2, respectively, as shown in Fig. 2.17. Choose from (a) through (d) the best statement about their motion.

(a) The ball thrown toward target 1 gets to its target earlier than the other ball does.

(b) The ball thrown toward target 2 gets to its target earlier than the other ball does.

(c) Both the balls get to their targets simultaneously.

(d) Which ball gets to its target earlier depends on the initial speed of the two balls.

(the 1st Challenge)

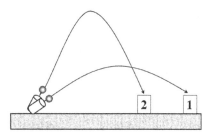

Fig. 2.17.

Answer (a)

Hint Consider the time taken for a ball to come back to the horizontal plane on which the ball is thrown.

Solution

Let v_0 be the initial speed and θ be the angle between the initial velocity and the horizontal plane. We take the vertical axis to be the y-axis. If the ball is shot from a point at $y = 0$ at $t = 0$, the y-coordinate of the ball at time t is

$$y = v_0 \sin \theta \cdot t - \frac{1}{2}gt^2.$$

Since the both targets are located on the same horizontal plane as the initial position, we set $y = 0$ in the preceding equation to obtain the time taken to reach the targets as

$$t = \frac{2v_0 \sin \theta}{g}.$$

This formula implies that the time taken decreases as the shooting angle, θ, decreases. ∎

Problem 2.4.   The motion of a train

The speed, v, of a train that travels between two stations (A and B) connected by a straight railway is plotted as a function of t in Fig. 2.18. The train starts from A at $t = 0$ with a constant acceleration of α. It slows down after $t = t_{\mathrm{m}}$ with a constant acceleration of $\beta(< 0)$, and at $t = T$, it stops at station B. Fill the boxes $\boxed{\mathrm{a}}$ through $\boxed{\mathrm{j}}$ in the following description about the motion of the train with the appropriate numbers or mathematical expressions.

Since the train is in uniform acceleration until $t = t_{\mathrm{m}}$, the velocity of the train before t_{m} can be written in terms of α and t as

$$v = \boxed{\mathrm{a}}. \tag{2.25}$$

To determine the time dependence of the velocity after $t = t_{\mathrm{m}}$, the velocity is written as

$$v = \beta t + c. \tag{2.26}$$

Since the values of v calculated by Eqs. (2.25) and (2.26) should be the same at $t = t_{\mathrm{m}}$, the constant c is

$$c = \boxed{\mathrm{b}} \times t_{\mathrm{m}}.$$

Hence, the velocity after $t = t_{\mathrm{m}}$ is

$$v = \beta(t - t_{\mathrm{m}}) + \boxed{\mathrm{c}}. \tag{2.27}$$

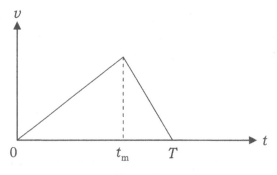

Fig. 2.18.

Using this expression, we can determine the time required for the train to arrive at B, T. From (2.27), we find in terms of α and β the relation between t_m and T as

$$t_m = \boxed{\text{d}} \times T. \tag{2.28}$$

Let s denote the distance between station A and the train. Derive an expression for s as a function of t. When t is less than t_m, s increases with t as follows:

$$s = \boxed{\text{e}}. \tag{2.29}$$

Whereas, the distance after $t = t_m$ is

$$s = \boxed{\text{f}} \times (t - t_m)^2 + \boxed{\text{g}} \times (t - t_m) + d, \tag{2.30}$$

where d is a constant determined below.

Since the values of s calculated from Eqs. (2.29) and (2.30) should be the same at $t = t_m$, the constant d is

$$d = \boxed{\text{h}}. $$

Let L denote the distance between the two stations. Then, we find in terms of α and β the relation between L and T^2 from Eqs. (2.28) and (2.30) as

$$\frac{L}{T^2} = \boxed{\text{i}}. \tag{2.31}$$

Suppose the distance between A and B is $L = 1.8\,\text{km}$ and the train travels with accelerations $\alpha = 0.20\,\text{m/s}^2$ and $\beta = -0.80\,\text{m/s}^2$, then, time T is $\boxed{\text{j}}$ seconds.

Answer

$$a = \alpha t, \quad b = \alpha - \beta, \quad c = \alpha t_m, \quad d = \frac{\beta}{\beta - \alpha}, \quad e = \frac{1}{2}\alpha t^2,$$

$$f = \frac{1}{2}\beta, \quad g = \alpha t_m, \quad h = \frac{1}{2}\alpha t_m^2, \quad i = \frac{\alpha\beta}{2(\beta - \alpha)}, \quad j = 150.$$

Hint Refer to Eqs. (2.1) and (2.2), which hold for linear motions each with a constant acceleration.

Solution

b: Since the values of v at $t = t_m$ evaluated by Eqs. (2.25) and (2.26) should be the same, we have

$$\alpha t_m = \beta t_m + c \quad \therefore c = \underline{(\alpha - \beta)t_m}.$$

c: Substituting the above expression for c into Eq. (2.26) yields

$$v = \beta t + (\alpha - \beta)t_m = \underline{\beta(t - t_m) + \alpha t_m}.$$

d: Since the train stops at $t = T$, it means $\alpha t_m + \beta(T - t_m) = 0$. It follows that

$$t_m = \underline{\frac{\beta}{\beta - \alpha}T}. \tag{2.28}$$

f, g: Since the velocity at $t = t_m$ is αt_m and the acceleration for $t > t_m$ is β, distance s is

$$s = \underline{\frac{1}{2}\beta \times (t - t_m)^2 + \alpha t_m(t - t_m) + d}, \tag{2.30}$$

where d is the distance between station A and the train at $t = t_m$.

i: After setting $t = T$ in Eq. (2.30), we have

$$L = \frac{1}{2}\beta(T - t_m)^2 + \alpha t_m(T - t_m) + \frac{1}{2}\alpha t_m^2.$$

Substituting Eq. (2.28) into this equation yields

$$\frac{L}{T^2} = \underline{\frac{\alpha\beta}{2(\beta - \alpha)}}. \tag{2.31}$$

j: Substituting the given values of α, β and L into Eq. (2.31) yields

$$T = \sqrt{\frac{2(\beta - \alpha)}{\alpha\beta}L} = \underline{150}\,\text{s}.$$

■

Problem 2.5. Skydiving

Suppose a group of individuals who are freely falling down performs a skydiving stunt. These skydivers jump off an aircraft individually and meet together in the sky. In this problem, we consider how a skydiver, A, catches up with another skydiver, B, who was initially below A. Notice that the air resistance on a body increases with its speed in air. For each of the following questions, choose the best answer from (a) through (d).

(1) Which of the following is the most appropriate description about the speed or the acceleration of a skydiver?

   (a) As soon as a skydiver leaves the aircraft, his speed is kept constant by air resistance.
   (b) The acceleration of a skydiver is kept constant by air resistance.
   (c) The acceleration of a skydiver increases during his fall whether air resistance acts on him.
   (d) The speed of fall approaches a terminal constant value due to air resistance.

(2) What should skydiver A do in order to catch up with skydiver B?

   (a) The higher skydiver A should hunch his body in order to increase the gravitational force acting on him.
   (b) The higher skydiver A should stretch out his body in order to increase the gravitational force acting on him.
   (c) The higher skydiver A should hunch his body in order to reduce the air resistance acting on him.
   (d) The higher skydiver A should stretch out his body in order to reduce the air resistance acting on him.

(3) After the performance, each skydiver returns to the ground with a parachute. What is the direction of his acceleration immediately after his parachute opens?

   (a) Upward
   (b) Downward

(c) Downward if his speed of fall is large enough

(d) No acceleration

(the 1st Challenge)

Answer (1) (d), (2) (c), (3) (a)

Hint Sometime after, the speed of a person who is falling in the presence of air resistance approaches a constant value called **the terminal velocity**. The terminal velocity increases with a decrease in air resistance, and it decreases with an increase in air resistance.

Solution

(1) A skydiver who is falling in the sky experiences an air resistance in the direction opposite to the velocity. Suppose the magnitude of the resistance is proportional to the speed of the skydiver, v, and the constant of proportionality is $k(>0)$. In terms of the mass of the skydiver, m, his downward acceleration, a, and the gravitational acceleration, g, the equation of motion is

$$ma = mg - kv. \tag{2.32}$$

(see Fig. 2.19). Since $v = 0$ initially, it follows from Eq. (2.32) that $a = g$. As the speed of fall increases, the acceleration decreases and approaches zero. As a result, the speed approaches the constant value $v_0 = \frac{mg}{k}$. After the skydiver leaves the aircraft at $t = 0$, his speed, v, changes with t as shown in Fig. 2.20.

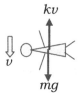

Fig. 2.19.

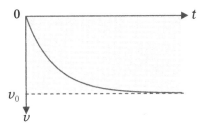

Fig. 2.20.

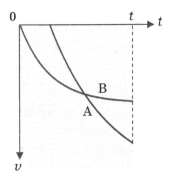

Fig. 2.21.

(2) If skydiver A changes his posture so that the air resistance he experiences decreases, the constant of proportionality, k, becomes smaller. Hence, the speed of skydiver A becomes larger than that of skydiver B as shown in Fig. 2.21, and consequently, skydiver A catches up to skydiver B at time t.

(3) Suppose a skydiver falls at a constant speed, v_0, when he opens his parachute at $t = t_0$. Immediately after, the air resistance he experiences increases abruptly and his acceleration can be calculated from an equation of motion in which the k in Eq. (2.32) is replaced by a larger constant, k_1. A long time after, the speed of the skydiver approaches a new terminal velocity $v_1 = \frac{mg}{k_1}(<v_0)$ as shown in Fig. 2.22, whose curve implies that after $t = t_0$, the acceleration of the skydiver becomes upward, and that his speed decreases. ∎

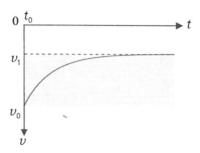

Fig. 2.22.

Supplement

(1) Substituting $a = \frac{dv}{dt}$ and $v_0 = \frac{mg}{k}$ into Eq. (2.23), we have

$$m\frac{dv}{dt} = -k(v - v_0).$$

This is an example of a differential equation that can be solved by separation of variables as follows.

We first remove the factor $v - v_0$ to the left-hand side and integrate both sides with respect to t:

$$\int \frac{1}{v - v_0}\frac{dv}{dt}\,dt = -\int \frac{k}{m}dt \Rightarrow \int \frac{dv}{v - v_0} = -\frac{k}{m}\int dt.$$

After performing the integration, we obtain

$$\log|v - v_0| = -\frac{k}{m}t + C,$$

where C is an arbitrary constant. Since the left-hand side of this equation diverges at $v = v_0$, the value of $v - v_0$ cannot be zero and thus cannot change its sign over time. When the motion begins with $v = 0$ at $t = 0$, v is smaller than $v_0 (v < v_0)$ forever. Hence, we have

$$v_0 - v = e^C \cdot e^{-\frac{k}{m}t}.$$

From the initial condition, we have $e^C = v_0$ and obtain

$$v = v_0(1 - e^{-\frac{k}{m}t}).$$

v is plotted as a function of t in Fig. 2.20.

(3) The skydiver opens his parachute at $t = t_0$ after the performance. The equation of motion of the skydiver for $t > t_0$ is

$$m\frac{dv}{dt} = mg - k_1 v = -k_1\left(v - \frac{mg}{k_1}\right) = -k_1(v - v_1).$$

The increase in the air resistance he experiences means $k_1 > k_0$ and, accordingly, $v_1 < v_0$. Since the new motion begins with the initial condition, $v = v_0 (> v_1), v - v_1$ is positive at all times. Hence, we have

$$v - v_1 = e^C \cdot e^{-\frac{k_1}{m}t}$$

From the initial condition $v = v_0$ at $t = t_0$, we have $e^C = (v_0 - v_1)e^{\frac{k_1}{m}t_0}$. Substituting this into the above equation, we obtain

$$v = v_1 + (v_0 - v_1)e^{-\frac{k_1}{m}(t-t_0)},$$

which holds for $t > t_0$. This equation implies that v tends to v_1 as t goes to infinity. v is plotted as a function of t in Fig. 2.22.

Problem 2.6. Small objects sliding on different descendent paths

Two paths (A and B) are shown in Fig. 2.23. The heights of the starting points of the two paths are the same, as are those of their ending points, and the horizontal distance, l, between the starting and ending points is the same for both paths. Each path involves different slopes. Suppose two identical objects are placed at the

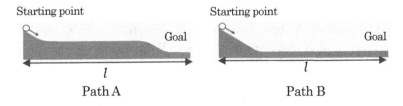

Fig. 2.23.

starting points of the two paths and begin to slide simultaneously. Choose from the following (a) through (d) the best statement about the motion of the two objects. Ignore any friction on the objects and assume that the objects are always in contact with the slopes.

(a) Since the object is accelerated twice in path A, the object in path A arrives at the goal earlier.
(b) After the object descends the first slope, the speed of the object in path B is larger than that in path A. Hence, the object in path B arrives at the goal earlier.
(c) Since both distances between the starting point and the goal are the same, the objects arrive at the goals simultaneously.
(d) Since the heights of the starting points and of the goals are the same in both paths A and B, the objects in both paths arrive at their goals simultaneously.

(the 1st Challenge)

Answer (b)

Hint The object that travels a longer distance at a faster speed arrives at the goal first.

Solution

An object gains a larger speed as the vertical distance of fall from the starting point becomes larger. Hence, the object that drops a longer distance gets a larger speed and, consequently, arrives at the goal first.   ∎

Problem 2.7.   An inclined plane

A ball is placed at the top of an inclined plane (of height h) as shown in Fig. 2.24, and it starts moving down. We consider the motion of the ball in the following two cases:

A. The ball slides down without rolling on a frictionless plane.
B. The ball rolls down without sliding on a plane with friction.

Fig. 2.24.

Choose from the following (a) through (d) the best statement about the time taken for the ball to reach the ground.

(a) The ball in case A reaches the ground in a shorter time than that in case B because the ball falls down without lossing mechanical energy.

(b) The ball in case B reaches the ground in a shorter time than that in case A because the ball falls down with rotational energy.

(c) Both balls reach the ground at the same time because of the law of conservation of energy. (The heights of the starting points in both cases are the same, as are the heights of the goals.)

(d) The ball in case A reaches the ground in a shorter time than that in case B because all of the potential energy of the ball is converted into kinetic energy due to its translational motion without being converted into rotational energy.

(the 1st Challenge)

Answer (d)

Hint When the ball rotates, part of its potential energy is converted into rotational energy.

Solution

When the ball falls down, its gravitational potential energy is converted into translational kinetic energy and rotational energy. If there is friction, the ball rotates; whereas in the frictionless case, the ball does not rotate. If there is no rotation, there is no rotational energy. So, in case A, all of the potential energy of the ball is converted into translational kinetic energy and the speed of the ball becomes larger. ■

Problem 2.8.   A space probe launched to converge with the orbit of Pluto

Celestial bodies that are farther than Neptune and are classified as dwarf planets have been found, in addition to Pluto. We consider a space probe that is launched to the neighborhood of the orbit of Pluto. In the following discussion, it should be noted that all objects, including planets, revolving around the Sun obey Kepler's laws.

(1) In accurate terms, the orbit of the Earth is an ellipse, but it is approximately a circle.

   The radius of this circle is the astronomical unit. Pluto's semi-major axis is about 40 astronomical units. What is Pluto's orbital period? Choose the best values from the following (a) through (f).

   (a) 10 years   (b) 40 years   (c) 180 years

   (d) 250 years   (e) 640 years   (f) 1600 years

(2) We launch the space probe at a speed larger than that of the Earth, in the direction tangent to the orbit of the Earth, as shown in Fig. 2.25. After being launched, the probe moves in an elliptical orbit such that its nearest point to the Sun is also a

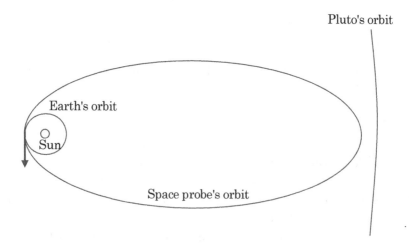

Fig. 2.25.

point on the orbit of the Earth. As shown in Fig. 2.25, if the semi-major axis of orbit of the space probe is about 20 astronomical units, its farthest point from the Sun in the orbit is near the orbit of Pluto. In this case, how many years does it take for this space probe to travel from the Earth to the neighborhood of the orbit of Pluto? Choose the best value from (a) through (f).

(a) 5 years (b) 30 years (c) 45 years
(d) 60 years (e) 90 years (f) 120 years

(3) Since the orbit of the space probe is an ellipse, the probe will come back to the neighborhood of the Earth after some time. Although the gravitational force due to the Sun, which decreases as the distance from the Sun increases, is very weak near Pluto's orbit, it can draw the space probe back to the neighborhood of the Earth. How can we explain the dynamics of this motion? Fill the following boxes $\boxed{a}$ through $\boxed{d}$ with the appropriate words, numbers or mathematical expressions.

If the space probe moves around the Sun at the same speed as that of the Earth, the gravitational force on the probe, F_G, balances with the centrifugal force, F_C. However, we launch the space probe at v_1, a speed larger than that of the Earth, in the direction tangent to the orbit of the Earth. Just after the launch, F_C exceeds $F_G (F_C > F_G)$, and consequently, the space probe moves away from the Sun.

We resolve the velocity vector of the space probe at a distance, r, from the Sun, $\vec{v} = \frac{d\vec{r}}{dt}$, into its radial component, v_r, and its component perpendicular to the radius, v_θ (Fig. 2.26). Kepler's second law can be written in terms of r and v_θ as

$$\boxed{a} = k \quad \text{(constant)}.$$

Now, by assuming the probe is in circular motion with a radius of r and with a speed of v_θ, we can describe the centrifugal force on the space probe of mass m as

$$F_C = m\frac{v_\theta^2}{r}.$$

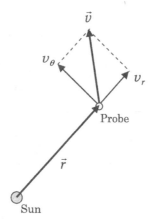

Fig. 2.26.

After eliminating v_θ from the expression for F_C, we obtain the centrifugal force in terms of k as

$$F_C = \boxed{\text{b}}.$$

Therefore, the magnitude of the centrifugal force is inversely proportional to the $\boxed{\text{c}}$ power of r.

Since the magnitude of the gravitational force is inversely proportional to the $\boxed{\text{d}}$ power of r, the centrifugal force decreases faster than the gravitational force as r increases.

When the centrifugal force becomes equal to the gravitational force ($F_C = F_G$), the radial component of the velocity, v_r, is still positive, that is, in the direction pointing away from the Sun, and hence, the space probe is moving away from the Sun. Immediately after, the centrifugal force becomes smaller than the gravitational force ($F_C < F_G$). Hence, the radial component of the acceleration becomes negative, and v_r decreases. If the initial velocity of the space probe, v_1, is not large enough, v_r vanishes in a while, and then, the space probe begins to move toward the Sun.

(4) If the initial velocity of the space probe, v_1, is large enough, the space probe escapes from the solar system. Let us discuss the space probe in that situation in terms of its mechanical energy.

When a space probe is at a distance, r, from the Sun, the potential energy associated with the gravitational force exerted by the Sun is

$$U = -G\frac{Mm}{r},$$

where M denotes the mass of the Sun and G denotes the gravitational constant. Hence, when the space probe moves around the Sun in the Earth's orbit, its mechanical energy is

$$E = \frac{1}{2}mv_0^2 - G\frac{Mm}{r_0}.$$

where v_0 is the orbital speed of the Earth and r_0 is the orbital radius of the Earth. How many times larger is the initial speed, v_1, required for the space probe to escape the solar system as compared with the speed v_0? Write down in detail how you derive your answer.

(5) There is a nest of comets known as "the Oort cloud" outside Pluto's orbit. A comet shot out from this nest comes close to the Sun but passes over the Sun without colliding it.

In the neighborhood of the Sun, the attractive gravitational force on the comet is very strong. Meanwhile, the comet has a lot of kinetic energy converted from its potential energy. By referring to the statements of (3), explain why the comet moves away from the Sun.

(the 1st Challenge)

Answer (1) (d), (2) (c)

(3) a $= \frac{1}{2}rv_\theta$ or rv_θ b $= \frac{4mk^2}{r^3}$ or $\frac{mk^2}{r^3}$ c $=$ 3rd   d $=$ 2nd

(4) As long as the mechanical energy of the space probe, E, is not negative, namely, $E \geq 0$, it can escape the solar system. When the space probe moves around the Sun in the Earth's orbit, its equation of motion

$$m\frac{v_0^2}{r_0} = G\frac{Mm}{r_0^2}. \tag{2.33}$$

On the other hand, when the space probe has an initial velocity of v_1, its mechanical energy is

$$E = \frac{1}{2}mv_1^2 - G\frac{Mm}{r_0}. \tag{2.34}$$

With Eqs. (2.33) and (2.34) and the condition $E \geq 0$,

$$E = \frac{1}{2}mv_1^2 - mv_0^2 \geq 0 \quad \therefore \frac{v_1}{v_0} \geq \sqrt{2} \text{ (times)}.$$

(5) The magnitude of the gravitational force is inversely proportional to the square of r, the magnitude of the radius vector, and the magnitude of centrifugal force is inversely proportional to the cube of r. Hence, when the comet comes close to the Sun, r becomes small, and consequently, the centrifugal force becomes larger than the gravitational force. Hence, the comet moves away from the Sun.

Solution

(1) Let the orbital radius of the Earth be $R_E = 1$ (astronomical unit), the orbital period of the Earth be $T_E = 1$ (year), the semi-major axis of Pluto's orbit be $R_P = 40$ (astronomical unit) and the orbital period of Pluto be T_P (year). Then, from Kepler's third law

$$\frac{T_P^2}{R_P^3} = \frac{T_E^2}{R_E^3} = 1,$$

we derive $T_P \approx 253$ (year).

(2) Let the semi-major axis of the space probe's orbit be $R_I = 20$ (astronomical unit) and the orbital period of the space probe be T_I (year). Then, from Kepler's third law, $\frac{T_I^2}{R_I^3} = 1$, and we derive $T_I \approx 89.4$ (year). Hence, the required time is $\frac{T_I}{2} = 44.7$ (year).

(3) As shown in Fig. 2.26, the motion of the space probe can be decomposed into a radial motion and a circular motion (of radius $r[= |\vec{r}|]$ and of speed v_θ). Note that a the centrifugal force of $m\frac{v_\theta^2}{r}$ and a gravitational force of $-G\frac{Mm}{r^2}$ (the negative sign implies the force is attractive) act on the space probe in the radial direction.

The centrifugal force can be derived by the formulation of the motion in terms of energies.

We take the reference point for defining the gravitational potential energy to be a point at infinity. The mechanical energy of the space probe, E, can be described as

$$E = \frac{1}{2}m(v_r^2 + v_\theta^2) - \frac{GMm}{r}.$$

From Kepler's second law, the relation between r and v_θ becomes

$$\frac{1}{2}rv_\theta = k = \text{constant.} \tag{2.35}$$

Hence, E can be described as

$$E = \frac{1}{2}mv_r^2 + U_e(r)$$

$$U_e(r) = \frac{2mk^2}{r^2} - \frac{GMm}{r}. \tag{2.36}$$

The total mechanical energy is expressed as the sum of the kinetic energy in the radial direction and the **effective potential energy**, $U_e(r)$. In general, a conservative force acting on an object points from a location of higher potential energy to one of lower potential energy, and its magnitude is given by the gradient of the potential energy. Thus, the force acting on the space probe in the radial direction, F, is

$$F = -\frac{dU_e}{dr} = \frac{4mk^2}{r^3} - \frac{GMm}{r^2}. \tag{2.37}$$

Now, using Eq. (2.35), we rewrite the rightmost term of Eq. (2.37) to obtain

$$F = \frac{mv_\theta^2}{r} - \frac{GMm}{r^2}.$$

This equation shows that the centrifugal force acting on the space probe in the direction of increasing r is $\frac{mv_\theta^2}{r} = \frac{4mk^2}{r^3}$ and the gravitational force acting on it in the direction of decreasing r is $\frac{GMm}{r^2}$.

We can comprehend the motion of the space probe by comparing the centrifugal force with the gravitational force.   ∎

Advanced Course

In the following advanced course, we write symbols representing vector in boldface italic type. Moreover, dotted symbols above a physical quantity represent its derivatives with respect to time. For example, we write $\vec{a}$ as $\boldsymbol{a}$, $\frac{dx}{dt}$ as $\dot{x}$, and $\frac{d^2x}{dt^2}$ as $\ddot{x}$.

2.5. Conservation of Momentum

2.5.1. *Momentum and Impulse*

Suppose a particle of mass m moves along the x-axis. Under the influence of a force, f, the equation of motion of the particle is

$$m\frac{dv}{dt} = f, \tag{2.38}$$

where $\frac{dv}{dt}$ is the acceleration of the particle. By integrating Eq. (2.38) with respect to time t from t_1 (at this time, the velocity of the particle is v_1) to t_2 (at this time, the velocity of the particle is v_2), we get

$$m\int_{t_1}^{t_2} \frac{dv}{dt}dt = \int_{t_1}^{t_2} f\,dt \;\Rightarrow\; m\int_{v_1}^{v_2} dv = \int_{t_1}^{t_2} f\,dt.$$

Furthermore, by replacing $\int_{t_1}^{t_2} f\,dt$ by I, we have

$$mv_2 - mv_1 = I. \tag{2.39}$$

Here, the product of the particle's mass and velocity, and I are called **momentum** and **impulse**, respectively. Equation (2.39) implies that

> **" the change in the momentum of a particle is**
>
> **equal to the applied impulse"**   (2.40)

2.5.2. *The Law of Conservation of Momentum*

Suppose particle 1 with a mass of m_1 and particle 2 with a mass of m_2 each exerts a force on the other; this pair of forces is of the same magnitude, f, (but is opposite in direction). Such forces are called **internal forces**. At time t_1, particles 1 and 2 have velocities v_1 and v_2, respectively, and at time t_2, a later time, they have velocities v_1' and v_2', respectively (Fig. 2.27). Suppose no forces act on the particles by the surroundings. (Forces acting on them by the surroundings are called **external forces**.) When an impulse, I, acts on particle 1, there is another impulse, $-I$, acting on particle 2 (this is called the law of **action-reaction**):

$$mv_1' - mv_1 = I, \quad mv_2' - mv_2 = -I.$$

The above equations yield

$$mv_1' + mv_2' = mv_1 + mv_2. \tag{2.41}$$

Equation (2.41) shows that the total momentum is constant. This relation is called the **law of conservation of momentum**.

If an external impulse, I', is applied to particles 1 and 2, which exert internal forces on each other, Eq. (2.41) have to be modified to

$$(mv_1' + mv_2') - (mv_1 + mv_2) = I'. \tag{2.42}$$

Equation (2.42) shows that

> **"the change in the total momentum of a system**
> **is equal to the impulse applied on the system**
> **by external forces"** (2.43)

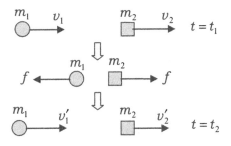

Fig. 2.27.

2.6. Moment of Force and Angular Momentum

As shown in Fig. 2.28, we represent the position of point P, an arbitrary point on a rigid body, by $r(|r| = r)$, a vector pointing from point O (a fixed point) to point P and called the position vector of point P. Suppose force, $f(|f| = f)$, acts on a point mass at P. Then, the **moment of force** about O, m, is defined as the following **vector product**:

$$m \stackrel{\text{def}}{=} r \times f. \qquad (2.44)$$

Therefore, the magnitude of m (the magnitude of the vector product $r \times f$) is equal to the area of the parallelogram formed by r and f, and it is given by $rf \sin \theta$. Here, θ ($0 \leq \theta \leq \pi$) is the angle between r and f. The moment of force, m, is perpendicular to the plane that contains the parallelogram, and is parallel to the direction in which a right-handed screw that rotates from the direction of r to that of f advances. If r is pointing in the direction of the x-axis and f is pointing in the direction of the y-axis, then m is pointing in the direction of the z-axis.

The magnitude of a vector product of two vectors A and B is maximum when A and B are perpendicular to each other ($A \perp B$), and it is zero when they are parallel ($A \| B$). If the components of A and B are given by $A = (A_x, A_y, A_z)$ and $B = (B_x, B_y, B_z)$,

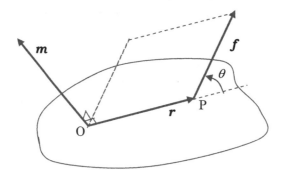

Fig. 2.28.

respectively, the vector product $\boldsymbol{A} \times \boldsymbol{B}$ is

$$\boldsymbol{A} \times \boldsymbol{B} = (A_y B_z - A_z B_y, A_z B_x - A_x B_z, A_x B_y - A_y B_x)$$

$$= \left(\begin{vmatrix} A_y & B_y \\ A_z & B_z \end{vmatrix}, \begin{vmatrix} A_z & B_z \\ A_x & B_x \end{vmatrix}, \begin{vmatrix} A_x & B_x \\ A_y & B_y \end{vmatrix} \right). \tag{2.45}$$

Here, $\begin{vmatrix} a & c \\ b & d \end{vmatrix} = ad - bc$ is called the **determinant** of the matrix $\begin{pmatrix} a & c \\ b & d \end{pmatrix}$.

Example 2.6. Derive the components of the vector product in Eq. (2.45) by using the distributive law for the vector product in terms of three vectors (namely, $\boldsymbol{a}, \boldsymbol{b}$ and $\boldsymbol{c}$):

$$(\boldsymbol{a} + \boldsymbol{b}) \times \boldsymbol{c} = \boldsymbol{a} \times \boldsymbol{c} + \boldsymbol{b} \times \boldsymbol{c}. \tag{2.46}$$

Solution

We define $\boldsymbol{i}, \boldsymbol{j}$ and $\boldsymbol{k}$ as the **unit vectors**, which are vectors each having a magnitude of unity, in the directions of the positive x-, y- and z-axes, respectively. Then, vectors $\boldsymbol{A} = (A_x, A_y, A_z)$ and $\boldsymbol{B} = (B_x, B_y, B_z)$ are represented as

$$\boldsymbol{A} = A_x \boldsymbol{i} + A_y \boldsymbol{j} + A_z \boldsymbol{k}, \quad \boldsymbol{B} = B_x \boldsymbol{i} + B_y \boldsymbol{j} + B_z \boldsymbol{k}.$$

Here, we use the distributive law (2.46) and the following equations about the vector products between the unit vectors:

$$\boldsymbol{i} \times \boldsymbol{i} = \boldsymbol{j} \times \boldsymbol{j} = \boldsymbol{k} \times \boldsymbol{k} = 0,$$
$$\boldsymbol{i} \times \boldsymbol{j} = \boldsymbol{k}, \quad \boldsymbol{j} \times \boldsymbol{k} = \boldsymbol{i}, \quad \boldsymbol{k} \times \boldsymbol{i} = \boldsymbol{j},$$
$$\boldsymbol{j} \times \boldsymbol{i} = -\boldsymbol{k}, \quad \boldsymbol{k} \times \boldsymbol{j} = -\boldsymbol{i}, \quad \boldsymbol{i} \times \boldsymbol{k} = -\boldsymbol{j}.$$

Then, Eq. (2.45) is derived as follows:

$$\boldsymbol{A} \times \boldsymbol{B} = (A_x \boldsymbol{i} + A_y \boldsymbol{j} + A_z \boldsymbol{k}) \times (B_x \boldsymbol{i} + B_y \boldsymbol{j} + B_z \boldsymbol{k})$$
$$= (A_y B_z - A_z B_y)\boldsymbol{i} + (A_z B_x - A_x B_z)\boldsymbol{j} + (A_x B_y - A_y B_x)\boldsymbol{k}.$$

∎

When a particle is at a position, r, and has a momentum, p, the **angular momentum** of the particle, l, is defined as

$$l \stackrel{\text{def}}{=} r \times p. \tag{2.47}$$

Here, l is a vector perpendicular to the plane containing r and p.

The derivative of the angular momentum, $\frac{dl}{dt}$, is equal to the moment of force, m:

$$\frac{dl}{dt} = m. \tag{2.48}$$

Example 2.7. Derive Eq. (2.48).

Solution

By differentiating Eq. (2.47), we obtain

$$\frac{dl}{dt} = \frac{dr}{dt} \times p + r \times \frac{dp}{dt}.$$

Since we can conclude $\frac{dr}{dt} \| p$ from $\frac{dr}{dt} = v$ and $p \propto v$, the first term in the right-hand side of the above equation is equal to zero. Furthermore, the second term becomes $r \times f = m$ because of the equation of motion $\frac{dp}{dt} = f$ (the derivative of $p = mv$). Hence, we get Eq. (2.48). ∎

2.7. The Keplerian Motion

We consider the motion of a body under the action of an universal gravitational force due to another body (this motion is called the **Keplerian motion**).

2.7.1. *Two-Dimensional Polar Coordinates*

We express the components of a two-dimensional position vector, r, in terms of r, the magnitude of r, and ϕ, the angle between r and the x-axis (Fig. 2.29). Then, (r, ϕ) is called the **two-dimensional polar coordinates** of r.

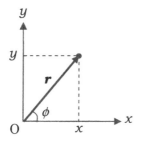

Fig. 2.29.

The relation between two-dimensional polar coordinates (r, ϕ) and two-dimensional rectangular coordinates (x, y) is

$$x = r \cos \phi, \quad y = r \sin \phi.$$

In order to represent the x-component and the y-component of a velocity vector, $\boldsymbol{v}$, where are denoted as v_x and v_y, respectively, in terms of two-dimensional polar coordinates, (r, ϕ), we differentiate the above equations:

$$v_x = \dot{x} = \dot{r} \cos \phi - r\dot{\phi} \sin \phi, \quad v_y = \dot{y} = \dot{r} \sin \phi + r\dot{\phi} \cos \phi. \quad (2.49)$$

We introduce the component of the velocity vector $\boldsymbol{v}$ in the direction of $\boldsymbol{r}$ (called the r-component of $\boldsymbol{v}$ and denoted as v_r) and the component perpendicular to $\boldsymbol{r}$ (called the ϕ-component of $\boldsymbol{v}$ and denoted as v_ϕ).

As shown in Fig. 2.30, the x- and y-components of $\boldsymbol{v}$ are expressed in terms of v_r and v_ϕ as

$$v_x = v_r \cos \phi - v_\phi \sin \phi, \quad v_y = v_r \sin \phi + v_\phi \cos \phi. \quad (2.50)$$

Comparing Eq. (2.49) with Eq. (2.50), we have

$$v_r = \dot{r}, \quad v_\phi = r\dot{\phi}. \quad (2.51)$$

Furthermore, the derivatives of Eq. (2.49) give the x- and y-components of the acceleration:

$$a_x = (\ddot{r} - r\dot{\phi}^2) \cos \phi - (2\dot{r}\dot{\phi} + r\ddot{\phi}) \sin \phi,$$
$$a_y = (\ddot{r} - r\dot{\phi}^2) \sin \phi + (2\dot{r}\dot{\phi} + r\ddot{\phi}) \cos \phi.$$

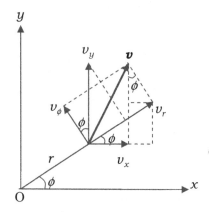

Fig. 2.30.

The following consideration is similar to the case of velocity. We write the r- and the ϕ-components of the acceleration as a_r and a_ϕ, respectively. Then, a_x and a_y are expressed in terms of a_r and a_ϕ as

$$a_x = a_r \cos \phi - a_\phi \sin \phi, \quad a_y = a_r \sin \phi + a_\phi \cos \phi. \quad (2.52)$$

Hence, we have

$$a_r = \ddot{r} - r\dot{\phi}^2, \quad a_\phi = 2\dot{r}\dot{\phi} + r\ddot{\phi}. \quad (2.53)$$

2.7.2. *Universal Gravitation Acting on Planets*

The mass of the Sun, M, is much larger than those of the other planets in the solar sytem, and so the Sun remains virtually at rest. The universal gravitational force, $\boldsymbol{F}$, exerted on a planet of mass m by the Sun is

$$\boldsymbol{F} = -\frac{GmM}{r^2}\frac{\boldsymbol{r}}{r},$$

where the vector $\boldsymbol{r}(|\boldsymbol{r}| = r)$ is the position vector pointing from the Sun to the planet. The negative sign above means that the force is attractive. G is the universal gravitational constant. The motion of equation is

$$m\ddot{\boldsymbol{r}} = -\frac{GmM}{r^2}\frac{\boldsymbol{r}}{r}. \quad (2.54)$$

2.7.3. *Moment of Central Forces*

A central force acting on a particle, f, is a force parallel to the position vector of the particle, r, namely, $(f\|r)$. The moment of a central force, m, is

$$m = r \times f = 0.$$

Then, $\frac{dl}{dt} = 0$ from Eq. (2.48), and so

$$l = \text{const.}$$

Therefore, **the angular momentum of a particle is conserved during the motion of the particle on which only central forces act.**

Example 2.8. Suppose the position of a particle of mass m on the $x - y$ plane is

$$r = (x, y, z) = (r \cos \phi, r \sin \phi, 0).$$

Here, r and ϕ are functions of time, t. Find l_z, the z-component of l, where l is the angular momentum of the particle.

Solution

The momentum of the particle, $p = (p_x, p_y, 0)$, is represented as

$$p_x = m\dot{x} = m(\dot{r}\cos\phi - r\dot{\phi}\sin\phi), \quad p_y = m\dot{y} = m(\dot{r}\sin\phi + r\dot{\phi}\cos\phi).$$

So, the z-component of the angular momentum is

$$
\begin{aligned}
l_z &= (r \times p)_z \\
&= x p_y - y p_x \\
&= r\cos\phi \cdot m(\dot{r}\sin\phi + r\dot{\phi}\cos\phi) - r\sin\phi \cdot m(\dot{r}\cos\phi - r\dot{\phi}\sin\phi) \\
&= mr^2\dot{\phi}.
\end{aligned}
\tag{2.55}
$$

■

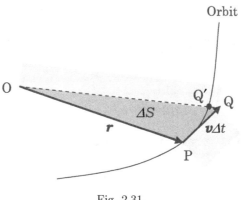

Fig. 2.31.

When a central force acts, we have

$$l_z = \text{const.}, \quad \text{i.e.,} \quad r^2\dot{\phi} = \text{const.}$$

Suppose a planet that experiences only the central forces moves on a curved orbit (see Fig. 2.31) at a velocity, $\boldsymbol{v}(|\boldsymbol{v}| = v)$. When the planet is displaced from point P to point Q′ in a small interval, Δt, the displacement $\overrightarrow{PQ'}$ is nearly equal to $\boldsymbol{v}\Delta t$ and the area of the sector OPQ′, ΔS, is approximated by the area of the triangle OPQ. Thus, the area ΔS is

$$\Delta S \approx \frac{1}{2}|\boldsymbol{r} \times \boldsymbol{v}\Delta t| = \frac{1}{2m}|\boldsymbol{r} \times \boldsymbol{p}|\Delta t = \frac{1}{2m}|\boldsymbol{l}|\Delta t.$$

Therefore, the areal velocity is

$$\frac{dS}{dt} = \frac{1}{2m}|\boldsymbol{l}|.$$

That is, the areal velocity is equal to the angular momentum divided by $2m$. For the motion of a planet under the influence of only central forces, we see that the areal velocity is constant, since $\boldsymbol{l}$ is constant.

Example 2.9. As shown in Fig. 2.32, the velocity of a particle changes from $\boldsymbol{v} = \overrightarrow{PQ}$ to $\boldsymbol{v'} = \overrightarrow{PQ'}$, as a central force $\boldsymbol{F}(\|\overrightarrow{OP})$,

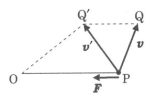

Fig. 2.32.

instantaneously acts on the particle. Then, show that the areal velocity about O is conserved.

Solution

Since the central force $\boldsymbol{F}$ is parallel to $\overrightarrow{OP}$, the velocity change $\overrightarrow{QQ'}$ is parallel to $\overrightarrow{OP}$:

$$\overrightarrow{QQ'} \parallel \overrightarrow{OP}.$$

Therefore, we have

$$\triangle OPQ = \triangle OPQ'.$$

This shows that the areal velocity of the particle is conserved when a central force acts on it. ∎

2.7.4. *Motion of Planets*

Since the universal gravitational force due to the Sun is a central force, the angular momenta of the planets around the Sun are conserved. Therefore,

$$L = mr^2\dot{\phi} = \text{const.}$$

From this result, we can derive the orbital equations of the planets using the two-dimensional polar coordinates.

Equation (2.54) can be written in terms of the x- and y-components of its terms as

$$m\ddot{x} = -\frac{GmM}{r^2}\frac{x}{r}, \quad m\ddot{y} = -\frac{GmM}{r^2}\frac{y}{r}.$$

By using $x = r\cos\phi$ and $y = r\sin\phi$ in the right-hand sides of the above equations, we obtain

$$\ddot{x} = -\frac{GM}{r^2}\frac{x}{r} = -\frac{GM}{r^2}\cos\phi, \tag{2.56}$$

$$\ddot{y} = -\frac{GM}{r^2}\frac{y}{r} = -\frac{GM}{r^2}\sin\phi. \tag{2.57}$$

Substituting $\frac{1}{r^2} = \frac{m}{L}\dot{\phi}$ into Eqs. (2.56) and (2.57) and replacing $\frac{GmM}{L}$ with μ yields

$$\ddot{x} = -\mu\dot{\phi}\cos\phi, \quad \ddot{y} = -\mu\dot{\phi}\sin\phi.$$

Then, by using $\frac{d}{dt}\sin\phi = \dot{\phi}\cos\phi$ and $\frac{d}{dt}\cos\phi = -\dot{\phi}\sin\phi$, we have

$$\ddot{x} = -\mu\frac{d}{dt}\sin\phi, \quad \ddot{y} = \mu\frac{d}{dt}\cos\phi.$$

After integrating the above equations over t, we get

$$\dot{x} = -\mu\sin\phi + C_1, \tag{2.58}$$

$$\dot{y} = \mu\cos\phi + C_2. \tag{2.59}$$

Here, C_1 and C_2 are integral constants.

Example 2.10. Using Eq. (2.49), derive an equation in terms of r, ϕ and $\dot{\phi}$ from Eqs. (2.58) and (2.59).

Solution

After substituting Eq. (2.49) into Eqs. (2.58) and (2.59), we have

$$\dot{r}\cos\phi - r\dot{\phi}\sin\phi = -\mu\sin\phi + C_1,$$
$$\dot{r}\sin\phi + r\dot{\phi}\cos\phi = \mu\cos\phi + C_2.$$

After summing the left-hand side of the first equation multiplied by $-\sin\phi$ and that of the second equation multiplied by $\cos\phi$, and equating it to the sum similarly obtained using the right-hand sides, we obtain

$$r\dot{\phi} = \mu - C_1\sin\phi + C_2\cos\phi. \tag{2.60}$$

■

To determine the two integral constants in Eq. (2.60), we consider the motion of a planet around the Sun. This motion is periodic. The angle ϕ changes from 0 to 2π. We define $\phi = 0$ $(r = r_1)$ at the perihelion, the point at which the planet is the closest to the Sun, and $\phi = \pi$ $(r = r_2)$ at the aphelion, the point at which the planet is the farthest from the Sun.

The conditions for the perihelion are

$$\phi = 0, \quad \frac{dr}{d\phi} = 0 \quad \text{and} \quad \frac{d^2 r}{d\phi^2} > 0 \quad \text{at } r = r_1.$$

Example 2.11. Determine C_1 in Eq. (2.60) from the conditions for the perihelion:

$$\phi = 0 \quad \text{and} \quad \frac{dr}{d\phi} = 0 \quad \text{at } r = r_1.$$

Solution

Substituting $\dot\phi = \frac{L}{mr^2}$ and $\mu = \frac{GmM}{L}$ into Eq. (2.60), we have

$$\frac{L}{mr} = \frac{GmM}{L} - C_1 \sin \phi + C_2 \cos \phi$$

$$\therefore \frac{1}{r} = \frac{Gm^2 M}{L^2} - \frac{mC_1}{L} \sin \phi + \frac{mC_2}{L} \cos \phi.$$

We differentiate the both sides of the above equation with respect to ϕ:

$$-\frac{1}{r^2} \frac{dr}{d\phi} = -\frac{mC_1}{L} \cos \phi - \frac{mC_2}{L} \sin \phi. \tag{2.61}$$

Then, we use the condition for the perihelion. The condition $\left(\frac{dr}{d\phi}\right)_{\phi=0} = 0$ gives

$$\frac{mC_1}{L} = 0 \quad \therefore \underline{C_1 = 0}. \qquad \blacksquare$$

Example 2.12. Find a bound on the possible values of the integral constant C_2 using the following conditions: $C_1 = 0$ and $\frac{d^2 r}{d\phi^2} > 0$

at $r = r_1$. Further, derive the following equation with appropriate positive constants D and ε:

$$r = \frac{D}{1 + \varepsilon \cos \phi}. \tag{2.62}$$

Solution

Since $C_1 = 0$, Eq. (2.61) can be simplified to $\frac{dr}{d\phi} = r^2 \frac{mC_2}{L} \sin \phi$. After differentiating this equation with respect to ϕ, we have

$$\frac{d^2 r}{d\phi^2} = 2r \frac{dr}{d\phi} \frac{mC_2}{L} \sin \phi + r^2 \frac{mC_2}{L} \cos \phi.$$

Using the condition for the perihelion $(\frac{d^2 r}{d\phi^2})_{\phi=0} > 0$ yields $r^2 \frac{mC_2}{L} > 0$. Since we can take $L > 0$ without loss of generality, we conclude $C_2 > 0$. Since $C_1 = 0$, the equation that relates r and ϕ becomes

$$\frac{1}{r} = \frac{Gm^2 M}{L^2} + \frac{mC_2}{L} \cos \phi = \frac{Gm^2 M}{L^2} \left(1 + \frac{LC_2}{GmM} \cos \phi \right).$$

Then, by defining $D = \frac{L^2}{Gm^2 M}$ and $\varepsilon = \frac{LC_2}{GmM}$, we obtain

$$r = \frac{D}{1 + \varepsilon \cos \phi}, \quad \varepsilon > 0. \qquad \blacksquare$$

Equation (2.62) is called **equation of the conic section in two-dimensional polar coordinates**. The conic sections are classified according to the value of ε (called **eccentricity**) as follows:

$$\begin{aligned}
\varepsilon &= 0 && \text{circle} \\
0 < \varepsilon &< 1 && \text{ellipse} \\
\varepsilon &= 1 && \text{parabola} \\
\varepsilon &> 1 && \text{hyperbola}
\end{aligned}$$

2.8. Motion and Energy of Rigid Bodies

A **rigid body** is an ideal body that has a perfectly unchanging shape and a definite size. The number of variables needed to specify the motion of a rigid body is called its **degrees of freedom**.

To specify the motion of a rigid body in space, we need the following six variables: the x-, y- and z-coordinates of its center of mass, the two angles θ and ϕ to determine the direction of the rotational axis that passes the center, and the rotational angle ψ. This means that the motion of a rigid body has six degrees of freedom.

2.8.1. *Motion of Rigid Bodies*

To study the motion of a body with six variables, we need six independent equations.

a. **The equation of translational motion** (contains three equations of components of vectors):

$$M\frac{d^2\boldsymbol{r}_{\mathrm{G}}}{dt^2} = \boldsymbol{F}, \quad (\boldsymbol{F} \text{ is the net external force}). \tag{2.63}$$

b. **The equation of rotational motion** (contains three equations of components of vectors):

$$\frac{d\boldsymbol{L}}{dt} = \boldsymbol{N}, \quad (\boldsymbol{N} \text{ is the net moment of force}). \tag{2.64}$$

Let us consider a rotational motion about a fixed axis.

This motion has only one degree of freedom. We may model a rigid body as a collection of a large number of particles of masses $m_i (i = 1, 2, \ldots)$. We express the angular velocity of a particle about the rotational axis as $\boldsymbol{\omega} = \dot{\psi}$ and the distance from the axis to the ith particle as r_i. The angular momentum of the rigid body is, then,

$$\boldsymbol{L} = \sum_i m_i r_i^2 \boldsymbol{\omega} = I\,\boldsymbol{\omega}. \tag{2.65}$$

Here, $I = \sum_i m_i r_i^2$ is called the **moment of inertia** of the rigid body about the rotational axis.

When we draw the z-axis along the fixed axis of the rotation and express the z-component of the moment of force as N_z, the equation of the rotational motion is

$$I\frac{d\omega}{dt} = N_z. \tag{2.66}$$

Example 2.13.

(1) Find I_1, the moment of inertia of a solid, uniform cylinder of mass M and radius a about its symmetry axis.
(2) Find I_2, the moment of inertia of a solid, uniform sphere of mass M and radius R about an axis through its center.

Solution

(1) We will, first, find the moment of inertia of a thin, uniform disk of mass m and radius a (Fig. 2.33), about an axis perpendicular to the disk and through its center. By using the mass per unit area $\sigma = \frac{m}{\pi a^2}$, the mass of the ring (of radius r and width dr), dm, is

$$dm = \sigma \cdot 2\pi r dr.$$

The moment of inertia of the disk, I, can, then, be calculated as

$$I = \int_0^a r^2 dm = 2\pi\sigma \int_0^a r^3 dr = 2\pi\sigma \left[\frac{r^4}{4}\right]_0^a = \frac{\pi\sigma}{2}a^4 = \frac{1}{2}ma^2.$$

We will, next, find I_1, the moment of inertia of the solid cylinder about its symmetry axis. As shown in Fig. 2.34, we divide the cylinder into thin disks each of mass $dM = m$ and

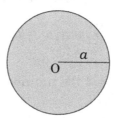

Fig. 2.33.

Axis

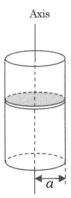

Fig. 2.34.

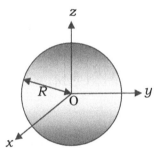

Fig. 2.35.

moment of inertia I. We integrate I to get I_1, the moment of inertia of the cylinder:

$$I_1 = \int \frac{1}{2} a^2 dM = \frac{1}{2} a^2 \int dM = \frac{1}{2} M a^2.$$

(2) We place the origin of the coordinate system at O, the center of the sphere, and position the x-, y- and z-axes as shown in Fig. 2.35. In terms of the density of mass (mass per unit volume) of the sphere, $\rho = \frac{M}{(4/3)\pi R^3}$, and the volume element (minute volume), dv, we can express the moments of inertia about the

x-, y- and z-axes as

$$I_x = \int \rho(y^2 + z^2)dv,$$

$$I_y = \int \rho(z^2 + x^2)\, dv, \quad I_z = \int \rho(x^2 + y^2)dv.$$

Here, $y^2 + z^2$, $z^2 + x^2$ and $x^2 + y^2$ are the squares of the distances from the position of a volume element, dv, to the x-, y- and z-axes, respectively. By considering the symmetry of the system, we get $I_2 = I_x = I_y = I_z$. Hence, we have

$$I_2 = \frac{1}{3}(I_x + I_y + I_z) = \frac{2}{3}\rho \int (x^2 + y^2 + z^2)\, dv = \frac{2}{3}\rho \int r^2 dv.$$

Finally, we substitute $4\pi r^2 dr$, the volume of a thin spherical shell of thickness dr, for dv to get I_2, the moment of inertial of the sphere:

$$I_2 = \frac{2}{3}\rho \int_0^R 4\pi r^4 dr = \frac{8}{3}\pi\rho \left[\frac{r^5}{5}\right]_0^R = \frac{8}{15}\pi\rho R^5 = \frac{2}{5}M R^2.$$

∎

2.8.2. *Rotational Kinetic Energy of Rigid Bodies*

We express the speed of the ith volume element rotating at an angular velocity, ω, as $r_i\omega$, where r_i is the distance from the axis to the element. Therefore, we can write the rotational kinetic energy of the element (of mass m_i) as $\frac{1}{2}m_i(r_i\omega)^2$.

The rotational kinetic energy of a rigid body, K_R, is the total sum of the kinetic energy of each of its element, and it is represented as $K_R = \sum_i \frac{1}{2}m_i(r_i\omega)^2$. It is expressed in terms of the moment of inertia of the body, I, and its angular velocity, ω, as

$$K_R = \frac{1}{2}I\omega^2. \tag{2.67}$$

Example 2.14. As shown in Fig. 2.36, a solid, uniform cylinder of mass M and radius R is initially ($t = 0$) at rest on a rough incline with a slope angle of θ. The central symmetry axis of the cylinder

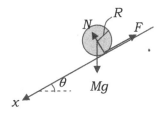

Fig. 2.36.

is horizontal. The coefficient of static friction, μ, the coefficient of kinetic (sliding) friction, μ', and the gravitational acceleration, g, are known.

(1) Find an inequality between μ and θ when the cylinder rolls down without sliding on the incline. And moreover, find the acceleration of the cylinder when it is rolling down on the incline.
(2) Consider the case that the cylinder rolls down with sliding on the incline. Find the velocity of the cylinder on the incline and the sliding velocity of the point at which the cylinder comes in contact with the incline as a function of time, t, while the cylinder descends.

Solution

(1) We choose the x-axis to point in the downward direction along the incline and express the coordinate of the symmetry axis of the cylinder as x. Since the cylinder rolls down without sliding, we write the equations of the translational and rotational motions in terms of F, the magnitude of the force of static friction on the cylinder, and I, the moment of inertia of the cylinder about its symmetry axis as

$$M\ddot{x} = Mg\sin\theta - F, \quad I\frac{d\omega}{dt} = FR. \qquad (2.68)$$

The relation between the velocity of the center of mass, $\dot{x}$, and the angular velocity, ω, is $\dot{x} = R\omega$. After substituting this equation and $I = \frac{1}{2}MR^2$ (see Example 2.12) into the second equation of

Eq. (2.68), we have

$$M\ddot{x} = 2F.$$

Comparing the above with the first equation of Eq. (2.68) gives $\ddot{x}$, the acceleration of the descending cylinder, and F, the force of the static friction, as

$$\ddot{x} = \frac{2}{3}g\sin\theta, \quad F = \frac{1}{3}Mg\sin\theta.$$

The magnitude of the normal reaction force is $N = Mg\cos\theta$. From $F \le \mu N$, we obtain the condition that the cylinder rolls down without sliding as

$$\tan\theta \le 3\mu.$$

(2) When the cylinder rolls down with sliding (i.e., $\tan\theta > 3\mu$), the magnitude of the force of kinetic friction is

$$F' = \mu'N = \mu' Mg\cos\theta.$$

Hence, the acceleration is as follows:

$$M\ddot{x} = Mg\sin\theta - \mu'Mg\cos\theta \Rightarrow \ddot{x} = g(\sin\theta - \mu'\cos\theta) > 0.$$

We obtain the velocity of the center of mass at time t as

$$\dot{x} = g(\sin\theta - \mu'\cos\theta)\,t,$$

after using the initial condition that $\dot{x} = 0$ when $t = 0$.

On the other hand, the equation of rotational motion about the center of mass is

$$I\frac{d\omega}{dt} = F'R = \mu' Mg\cos\theta \cdot R.$$

After substituting $I = \frac{1}{2}M R^2$ into the above equation, we get

$$R\dot{\omega} = 2\mu'g\cos\theta.$$

We, then, obtain the sliding velocity, u, as

$$u = \dot{x} - R\omega = g(\sin\theta - 3\mu'\cos\theta)\,t(>0),$$

after using the initial condition that $\omega = 0$ when $t = 0$. ∎

Advanced Problems

Problem 2.9.   The Atwood machine with friction

I Let us consider a model of the Atwood machine, which consists
of a square prism, ABCD, whose central axis is horizontal and
perpendicular to this sheet of paper and whose face AB makes an
angle of $\frac{\pi}{4}$ with the vertical line. On the upper faces AB and BC
of the prism, there are massless bodies P and Q, respectively. As
shown in Fig. 2.37, the two bodies P and Q are connected by an
inextensible massless string, S_1, and similar strings S_0 and S_2 are
hung down from P and Q, respectively. Note that the string S_1
is horizontal and strings S_0 and S_2 are vertical. Here, we assume
there is static friction between body P and face AB as well as
between body Q and face BC; the frictional coefficients for both
cases are μ_0. Suppose bodies P and Q are always in contact with
faces AB and BC, respectively.

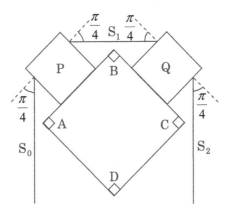

Fig. 2.37.

(1) Let the tension in string S_0 be T_0. Find T_1, the minimum tension in string S_1 to keep body P at rest.

(2) We hang W_0, a weight of mass M_0, at the lower end of string S_0, and W_2, another weight, at the lower end of string S_2. Find m_2, the minimum mass of W_2 needed to keep bodies P and Q at rest.

II We replace the square prism of Section I with an equilateral $2n$-sided prism ($n = 2, 3, \ldots$), and put n massless bodies on the n upper faces of the $2n$-sided prism. They are connected by inextensible massless strings as in Section I. All the strings are taut. The angle between each face of the $2n$-sided prism and its corresponding string is $\frac{\pi}{2n}$. The coefficient of static friction between each body and its corresponding face of the $2n$-sided prism is assumed to be equal to that in Section I, μ_0.

(3) We hang a weight of mass M from the leftmost body, and another weight of mass m from the rightmost body. Find the minimum value of mass m to keep all bodies at rest.

III A massless string is wound on a fixed cylinder that is unable to rotate. Its central axis is horizontal and perpendicular to this sheet of paper. A small body of mass $10\,\mathrm{g}$ is hung at the right end of the string, while a man of mass $60\,\mathrm{kg}$ is hung at the left end of the string (Fig. 2.38). Assume that the coefficient of static friction between the string and the face of the cylinder is 1.0.

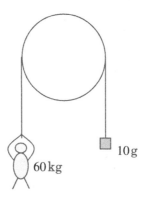

Fig. 2.38.

(4) How many times at least do we need to wind the string around the cylinder to keep the man hanging at rest? Suppose we define the number of turns as zero when the string is just laid over the cylinder, and suppose friction acts only between the string and the face of the cylinder but does not act between two wound parts of the string that are in contact. Here, one can use the following approximations in the case where x is much smaller than unity ($|x| \ll 1$),

$$\cos x \approx 1, \quad \sin x \approx x, \quad \frac{1}{1-x} \approx 1+x,$$

and in the case $h \to \infty$,

$$\left(1 + \frac{1}{h}\right)^h \to e \overset{\text{def}}{=} 2.718 \ldots$$

(e is the base of the natural logarithm.)

IV Suppose a massless string is just hung over a cylinder as in Section III (the number of turns in this situation is zero). Weight A, a weight of mass M_A, is connect to the left end of the string, and weight B, a weight of mass M_B ($<M_A$), is connected to the right end (Fig. 2.39). We pull B downward at an initial speed of v_0 from point O. Then, B moves down to the lowest point where it turns around, and starts moving upward. Then, it passes the original point O. Assume the coefficient of static friction between the string and the face of the cylinder is 1.0, and the coefficient of kinetic friction between them is $\mu(< 1.0)$.

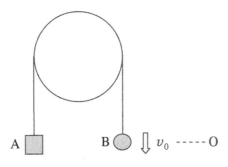

Fig. 2.39.

(5) Find the minimum value of the mass ratio $\frac{M_A}{M_B}$ in order for B to turn around at the lowest point. Express your result in two significant figures. ·

(6) Derive the formula for v_1, the speed of B at the instant when B returns to point O, and calculate the ratio $\frac{v_0}{v_1}$ to two significant figures when $\frac{M_A}{M_B} = 50$ and $\mu = 0.8$.

(the 1st Challenge)

Solution

(1) Let the normal force acting on body P be R, then the maximum value of the static frictional force is $\mu_0 R$. When the tension in string S_1 is T_1, the conditions for translational equilibrium are

$$T_0 \cos \frac{\pi}{4} = T_1 \cos \frac{\pi}{4} + \mu_0 R,$$

$$T_0 \sin \frac{\pi}{4} + T_1 \sin \frac{\pi}{4} = R,$$

in the directions parallel and normal to face AB, respectively. After eliminating R from these equations, we obtain

$$T_1 = \frac{\cos \frac{\pi}{4} - \mu_0 \sin \frac{\pi}{4}}{\cos \frac{\pi}{4} + \mu_0 \sin \frac{\pi}{4}} T_0 = \frac{1 - \mu_0}{1 + \mu_0} T_0.$$

(2) When the mass of weight W_2 is at its minimum value, m_2, both of the static frictional forces on bodies P and Q are at their maximum values. Let the gravitational acceleration be g. Then, in the same way as part (1), we obtain the tension $m_2 g$ in string S_2 as

$$m_2 g = \frac{\cos \frac{\pi}{4} - \mu_0 \sin \frac{\pi}{4}}{\cos \frac{\pi}{4} + \mu_0 \sin \frac{\pi}{4}} T_1.$$

Since $T_0 = M_0 g$, we have

$$m_2 = \left(\frac{\cos \frac{\pi}{4} - \mu_0 \sin \frac{\pi}{4}}{\cos \frac{\pi}{4} + \mu_0 \sin \frac{\pi}{4}} \right)^2 M_0 = \left(\frac{1 - \mu_0}{1 + \mu_0} \right)^2 M_0.$$

(3) When mass m is at its minimum value, the static frictional force acting on each body is at its a maximum value. Then, we have

$$m = \left(\frac{\cos \frac{\pi}{2n} - \mu_0 \sin \frac{\pi}{2n}}{\cos \frac{\pi}{2n} + \mu_0 \sin \frac{\pi}{2n}} \right)^n M.$$

(4) In the limit of $n \to \infty$, we can regard an equilateral $2n$-sided prism as a cylinder. Using the result of part (3), we have in the limit of $\frac{\pi}{2n} \to 0$.

$$M = \left(\frac{\cos \frac{\pi}{2n} + \mu_0 \sin \frac{\pi}{2n}}{\cos \frac{\pi}{2n} - \mu_0 \sin \frac{\pi}{2n}} \right)^n m \approx \left(\frac{1 + \mu_0 \frac{\pi}{2n}}{1 - \mu_0 \frac{\pi}{2n}} \right)^n m$$

$$\approx \left(1 + \mu_0 \frac{\pi}{n} \right)^n m.$$

If we define $h \overset{\text{def}}{=} \frac{n}{\mu_0 \pi}$, then we obtain in the limit of $h \to \infty$

$$\left(1 + \mu_0 \frac{\pi}{n} \right)^n = \left(1 + \frac{1}{h} \right)^{\mu_0 \pi h} \to e^{\mu_0 \pi}.$$

When the number of turns is zero, the minimum value of mass m needed to keep mass M at rest is given by

$$M = e^{\mu_0 \pi} m. \tag{2.69}$$

When the string is wound around the cylinder, we can treat the string tension on both lower and upper parts of the cylinder in the same way since the string is massless. Therefore, when the number of turns is N, the minimum value of mass m_0 needed to keep mass M at rest is given by

$$M = e^{(2N+1)\mu_0 \pi} m_0 \quad \therefore \quad N = \frac{1}{2} \left(\frac{1}{\mu_0 \pi} \ln \frac{M}{m_0} - 1 \right),$$

where $\ln$ is the natural logarithm whose base is e. After substituting $\mu_0 = 1.0$, $M = 60\,\text{kg}$, and $m_0 = 10 \times 10^{-3}\,\text{kg}$, we obtain

$$N \approx 0.88.$$

Therefore, the minimum number of turns is $\underline{1}$.

(5) In the same way as the derivation of Eq. (2.69), the condition for the weight B to turn around at the lowest point is

$$M_A > e^{\mu_0 \pi} M_B \quad \therefore \quad \frac{M_A}{M_B} > e^{\mu_0 \pi} = e^{\pi} \approx \underline{23}.$$

(6) When the string slides on the cylinder with an acceleration, the forces exerted on an infinitesimal part of the string on the cylinder must always be kept in balance because the string is massless. Therefore, the relation between T_A, the tension on weight A, and T_B, the tension on weight B, can be derived in the same way as the derivation of Eq. (2.69).

When B is falling, we have $T_A = e^{-\mu\pi}T_B$, whereas when B is rising, we have $T_A = e^{\mu\pi}T_B$.

When B is falling (Fig. 2.40), the equations of motion of A and B can be written as

$$M_A \alpha_d = M_A g - e^{-\mu\pi} T_B,$$
$$M_B \alpha_d = T_B - M_B g,$$

where α_d is the upward acceleration of B.

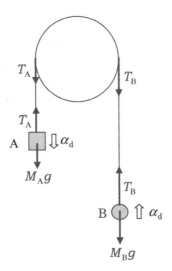

Fig. 2.40.

After eliminating T_B from these equations, we have

$$\alpha_d = \frac{M_A - M_B e^{-\mu\pi}}{M_A + M_B e^{-\mu\pi}} g = \text{const.}$$

Hence, by using the equation for a motion with a constant acceleration, we find L, the distance between point O and the lowest point of B, as follows:

$$0^2 - v_0^2 = 2(-\alpha_d)L,$$

$$\therefore L = \frac{v_0^2}{2\alpha_d} = \frac{1}{2}\left(\frac{M_A + M_B e^{-\mu\pi}}{M_A - M_B e^{-\mu\pi}}\right)\frac{v_0^2}{g}.$$

When B is rising, its upward acceleration, α_u, is obtained by replacing $e^{-\mu\pi}$ with $e^{\mu\pi}$ in the expression for α_d. That is,

$$\alpha_u = \frac{M_A - M_B e^{\mu\pi}}{M_A + M_B e^{\mu\pi}} g.$$

Hence, v_1, the speed of weight B passing point O, is obtained as follows:

$$v_1^2 - 0^2 = 2\alpha_u L$$

$$\therefore v_1 = \sqrt{2\alpha_u L} = v_0\sqrt{\frac{(M_A - M_B e^{\mu\pi})(M_A + M_B e^{-\mu\pi})}{(M_A + M_B e^{\mu\pi})(M_A - M_B e^{-\mu\pi})}}.$$

Then, we have

$$\frac{v_0}{v_1} = \sqrt{\frac{(M_A + M_B e^{\mu\pi})(M_A - M_B e^{-\mu\pi})}{(M_A - M_B e^{\mu\pi})(M_A + M_B e^{-\mu\pi})}}$$

$$= \sqrt{\frac{\left(\frac{M_A}{M_B} + e^{\mu\pi}\right)\left(\frac{M_A}{M_B} - e^{-\mu\pi}\right)}{\left(\frac{M_A}{M_B} - e^{\mu\pi}\right)\left(\frac{M_A}{M_B} + e^{-\mu\pi}\right)}} \approx \underline{1.3}.$$

Problem 2.10.   The rotation of rods

Consider a thin, uniform rod of length l and mass M. The moment of inertia of the rod about the transverse axis passing through its center of mass is given by $\frac{1}{12}Ml^2$.

I There are two thin, uniform rods (A and B) of equal length, l, and equal mass, M. Each rod is constrained at one of its end along a horizontal rail and is free to rotate about its constrained end in the vertical plane containing the horizontal rail. The constrained end of rod A can move without friction along the rail, whereas that of rod B is fixed to a point on the rail. Denote the gravitational acceleration as g.

  (1) Each rod is initially kept in the horizontal position, as shown in Fig. 2.41, and then quietly released. Find the ratio $\frac{\omega_A}{\omega_B}$ where ω_A and ω_B are the angular velocities of rods A and B, respectively, as a function of θ, the rod inclines the angle to the vertical line, as shown in Fig. 2.42.

  (2) Find the ratio $\frac{\omega_A}{\omega_B}$ at the instance each rod is just vertical (that is, $\theta = 0$).

  (3) Find the ratio $\frac{T_A}{T_B}$ where T_A and T_B are the periods of the sufficiently small oscillations of rods A and B about their vertical positions (that is, $\theta = 0$), respectively.

II Two thin, uniform rods (A and B) of equal length l and equal mass M are connected to each other without friction and are

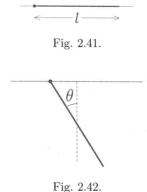

$$\longleftarrow l \longrightarrow$$

Fig. 2.41.

Fig. 2.42.

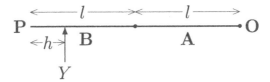

Fig. 2.43.

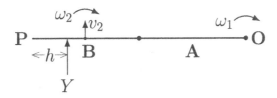

Fig. 2.44.

placed in a straight line on a horizontal smooth plane as shown in Fig. 2.43. The right end of rod A is constrained to a fixed point O, about which the rod can freely rotate without friction. A blow is struck at a point a distance $h(h < l)$ from, P, the left end of rod B. The direction of the blow is perpendicular to the line $\overline{OP}$. The impulse of the blow is denoted as Y.

(1) Just after the blow, find the angular velocity of rod A, ω_1, that of rod B, ω_2, and the velocity of the center of mass of rod B, v_2 (Fig. 2.44).

(2) Find the impulse delivered to the right end of rod A at the instant the blow is struck.

(3) Suppose after the blow, the rods begin to rotate and remain aligned in a straight line around the fixed point O. Find the distance h (the point where the blow is struck), and ω_0, the angular velocity of the rotation.

(the Final Challenge)

Solution

I (1) The case of rod A: Let the x-axis be the axis containing the horizontal rail and the y-axis be the vertical line passing through the center of the rod. Because there is no force acting in the direction of the x-axis, the center of mass of the rod

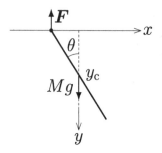

Fig. 2.45.

only moves along the y-axis. The y-coordinate of the center of mass of the rod, y_c, is $y_c = \frac{l}{2}\cos\theta$ (Fig. 2.45). Let I_0 be the moment of inertia about the transverse axis passing through the center of mass of the rod. Then, by the law of conservation of mechanical energy, we have

$$\frac{1}{2}M\dot{y}_c^2 + \frac{1}{2}I_0\dot{\theta}^2 = Mgy_c = Mg\frac{l}{2}\cos\theta,$$

where the moment of inertia I_0 is $I_0 = \frac{1}{12}Ml^2$. Using the relation $\dot{y}_c = -\frac{l}{2}(\sin\theta)\dot{\theta}$, we obtain

$$\frac{1}{24}Ml^2(1 + 3\sin^2\theta)\dot{\theta}^2 = \frac{1}{2}Mgl\cos\theta.$$

The angular velocity of the rod is, then,

$$\omega_A = |\dot{\theta}| = 2\sqrt{\frac{3\cos\theta}{1 + 3\sin^2\theta}\frac{g}{l}}.$$

The case of rod B: Let I be the moment of inertia about the transverse axis passing through the fixed end of the rod (Fig. 2.46). Then, by the law of conservation of mechanical energy, we have

$$\frac{1}{2}I\dot{\theta}^2 = Mg\frac{l}{2}\cos\theta,$$

where the moment of inertia I is

$$I = \frac{1}{12}Ml^2 + M\left(\frac{l}{2}\right)^2 = \frac{1}{3}Ml^2.$$

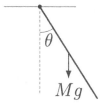

Fig. 2.46.

The angular velocity of the rod is, then,

$$\omega_{\mathrm{B}} = |\dot{\theta}| = \sqrt{\frac{3g\cos\theta}{l}}.$$

Hence, the ratio of the angular velocities is

$$\frac{\omega_{\mathrm{A}}}{\omega_{\mathrm{B}}} = \frac{2}{\sqrt{1+3\sin^2\theta}}.$$

(2) Substituting $\theta = 0$ into the above equation yields $\frac{\omega_{\mathrm{A}}}{\omega_{\mathrm{B}}} = 2$.

(3) Let us find the period of sufficiently small oscillations of each rod.

The case of rod A: Let F be the normal reaction force exerted upward on the upper end of the rod (Fig. 2.45). The equation of the linear motion of the center of mass and the equation of the rotational motion about the transverse axis passing through the center of mass are, respectively,

$$M\ddot{y}_{\mathrm{c}} = Mg - F,$$

$$I_0\ddot{\theta} = -\frac{1}{2}lF\sin\theta.$$

For sufficiently small oscillations about the equilibrium position, the approximate relation $\sin\theta \approx \theta$ can be used. Also, the variation in the y-coordinate of the center of mass is negligible, and so $F \approx Mg$. Then, the equation of the rotational motion yields

$$\ddot{\theta} = -\frac{(1/2)lMg}{(1/12)Ml^2}\theta = -\frac{6g}{l}\theta.$$

The angular frequency and the period of the oscillation are, respectively,

$$\omega = \sqrt{\frac{6g}{l}},$$

$$T_A = \frac{2\pi}{\omega} = 2\pi\sqrt{\frac{l}{6g}}.$$

The case of rod B: The equation of the rotational motion about the transverse axis passing through the fixed point is

$$I\ddot{\theta} = -Mg\frac{l}{2}\sin\theta.$$

For sufficiently small oscillations,

$$\ddot{\theta} = -\frac{3g}{2l}\theta.$$

The angular frequency and the period of the oscillation are, respectively,

$$\omega = \sqrt{\frac{3g}{2l}},$$

$$T_B = \frac{2\pi}{\omega} = 2\pi\sqrt{\frac{2l}{3g}}.$$

Accordingly, the ratio of the periods is

$$\frac{T_A}{T_B} = \frac{1}{2}.$$

II (1) At the instant the blow is struck, impulsive forces are exerted on the rods at the connection point as well as at the fixed point O in the direction perpendicular to the rods (Fig. 2.47). Let Y_o be the magnitude of the impulse of the force exerted on rod A at the fixed point O, and let Y_c be the magnitude of the impulse of the force exerted on rod A by rod B at the connection point. Then, Y_c is also the magnitude of the impulse of the reaction force exerted on rod B by rod A at the connection point.

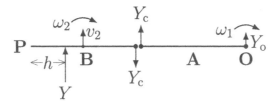

Fig. 2.47.

The change in the angular momentum of rod A about the fixed point O is equal to the moment of the impulse Y_c:

$$I\omega_1 = lY_c, \quad I = \frac{1}{3}Ml^2. \tag{2.70}$$

The change in the momentum of rod B is equal to the impulse $Y - Y_c$, and the change in the angular momentum of rod B about the transverse axis passing through its center of mass is equal to the sum of the moments of the impulses Y and Y_c. Thus we have

$$Mv_2 = Y - Y_c, \tag{2.71}$$

$$I_0\omega_2 = \left(\frac{l}{2} - h\right)Y + \frac{l}{2}Y_c, \quad I_0 = \frac{1}{12}Ml^2. \tag{2.72}$$

Note that the velocity of the connection point of rod A is equal to the velocity of the connection point of rod B:

$$l\omega_1 = v_2 - \frac{l}{2}\omega_2. \tag{2.73}$$

Substituting the expressions for ω_1, v_2 and ω_2 obtained in Eqs. (2.70) through (2.72) into Eq. (2.73) yields

$$l\frac{lY_c}{(1/3)Ml^2} = \frac{Y - Y_c}{M} - \frac{l}{2}\frac{(l/2 - h)Y + (l/2)Y_c}{(1/12)Ml^2}.$$

By solving this equation for Y_c, we obtain

$$Y_c = \frac{2(3h - l)}{7l}Y.$$

By substituting this relation into Eqs. (2.70) through (2.72), we obtain the following relations:

$$\omega_1 = \frac{6(3h - l)}{7l^2} \frac{Y}{M},$$

$$\omega_2 = \frac{6(5l - 8h)}{7l^2} \frac{Y}{M}, \quad v_2 = \frac{3(3l - 2h)}{7l} \frac{Y}{M}.$$

(2) The motion of rod A immediately after the blow is given by the combination of the translation of and the rotation about its center of mass. Let v_1 be the velocity of the center of mass of rod A just after the blow. Then, the change in the linear momentum of rod A is

$$M v_1 = Y_{\rm o} + Y_{\rm c},$$

and the change in its angular momentum is

$$I_0 \omega_1 = \frac{l}{2}(Y_{\rm c} - Y_{\rm o}).$$

Either equation gives the same result. Here, by using the former, we have

$$Y_{\rm o} = M v_1 - Y_{\rm c} = M \frac{l}{2} \omega_1 - Y_{\rm c}$$

$$= \frac{l}{2} \frac{6(3h - l)}{7l^2} Y - \frac{2(3h - l)}{7l} Y = \frac{3h - l}{7l} Y.$$

(3) If the two angular velocities ω_1 and ω_2 are the same, the two rods will rotate at the same angular velocity and remain aligned in a straight line. The relation $\omega_1 = \omega_2$ gives $h = \frac{6}{11}l$, and substituting this expression for h into the equations for ω_1 or ω_2 yields

$$\omega_1 = \omega_2 = \frac{6Y}{11Ml} = \omega_0.$$

Suppose the fixed point O is able to support the centrifugal force due to the circular motion of the rods, they will rotate about O without translation at a constant angular velocity of ω_0. ∎

Problem 2.11.   The expanding Universe

Using Newton's law of universal gravitation, we'll study the development of the Universe without taking into account the general theory of relativity.

When we observe lights coming from distant galaxies, the wavelength of the observed lights are shifted longer. The wavelength of the light emitted from a receding source is longer than that from a stationary source (or that observed in the rest frame of the source). This shift of the wavelength is called the **red shift**. The speed of light in vacuum is denoted as c.

(1) An atom receding at a speed, v, from an observer, O, emits light of a wavelength that has a value of λ_0 when observed in the rest frame of the atom. Express λ, the wavelength of the light observed by O, in terms of λ_0, c and v. Suppose v is much smaller than c ($v \ll c$). Therefore, it is not necessary to take account of the special relativity.

In 1929, Edwin Hubble, an American astronomer, found by observing the wavelengths of the lights coming from distant galaxies that most galaxies are receding from the Earth at speeds nearly proportional to their distances from the Earth. The relation between v, the receding speed of a galaxy, and r, its distance from the Earth, is

$$v = H_0 r.$$

Here, H_0 is called the **Hubble constant**. If this equation is valid for an arbitrary duration of t, the above relation can be written in terms of the receding speed $v(t)$, the distance $r(t)$ and a parameter $H(t)$ as

$$v(t) = H(t)r(t), \tag{2.74}$$

where $v(t) = \frac{dr(t)}{dt}$ is the receding speed and $H(t)$ is called the **Hubble parameter**, which is a function of time.

The Universe has neither special places nor special directions. This is called the **cosmological principle**. With this principle, we consider the expansion of a uniform and isotropic universe. Imagine its expansion as an inflation of a balloon. The distance between two arbitrary points on the surface of the balloon increases uniformly with the inflation. We can liken this increase of the distance on the surface (which is a two-dimensional space) to the increase of the distance between two galaxies in the Universe (which is a three-dimensional space).

We introduce, in terms of r_0, the present $(t = t_0)$ distance between an observer, O, and a particle, P, and $r(t)$, the distance at time t, a quantity called the **scale factor**, $a(t)$, which represents the scale of the expansion of the Universe:

$$a(t) = \frac{r(t)}{r_0}. \tag{2.75}$$

When an immense explosion called the **Big Bang** happened $(t = 0)$, we consider observer O and particle P to be at the same point. That is, we suppose $r(0) = 0$ and $a(0) = 0$. In the following, we simply write $r(t)$ as r and $a(t)$ as a.

(2) Represent the Hubble parameter, $H(t)$, in terms of the scale factor a and its derivative $\frac{da}{dt}$.

We denote the density of mass of a sphere (whose center is at a point O) with a radius of r_0 at time t_0 as ρ_0 and that of a sphere with a radius of r at time t as ρ. Suppose the total mass of a sphere is conserved as it expands.

When the mass is uniformly distributed, it seems to an observer at point O that the mass distribution is spherically symmetric around him. We denote the distance from a particle, P, to point O as r. Then, the net gravitational force on particle P due to the mass outside the sphere with a radius of r and centered at point O is zero. Hence, the net gravitational force on P is the same as that when the mass inside the sphere were all concentrated at O (Fig. 2.48). In the following, we denote the gravitational constant as G.

(3) When we write the acceleration of a particle P at a point of a distance r from point O as $\frac{d^2r}{dt^2}$, we have

$$\frac{d^2a}{dt^2} = \frac{1}{r_0}\frac{d^2r}{dt^2}.$$

Express $\frac{1}{a}\frac{d^2a}{dt^2}$ in terms of ρ and G.

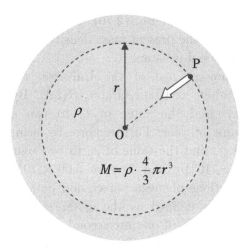

Fig. 2.48.

At present, the Universe is expanding. So, the possible future of the Universe is as follows:

(i) The Universe keeps expanding forever.
(ii) The Universe shrinks eventually.

Recent observations show that the fate of the Universe lies on a border between (i) and (ii). This means that the total mechanical energy of the Universe is nearly zero. In the following calculations, assume that the mechanical energy is zero.

(4) Suppose particle P recedes from point O. Express $(\frac{1}{a}\frac{da}{dt})^2$ in terms of ρ and G by using the law of conservation of mechanical energy of P.
(5) Suppose the total mass of a sphere is conserved as it expands. Express ρ in terms of ρ_0 and a.

From parts (4) and (5), we can see that the scale factor $a(t)$ at time t can be expressed as

$$a(t) = \left(\frac{t}{t_0}\right)^n. \tag{2.76}$$

Here, t_0 is the present time and we have $a(t_0) = 1$.

(6) Find n, the exponent in Eq. (2.76). Furthermore, express the time t_0 in terms of the present mass density of the Universe, ρ_0, and the gravitational constant, G.

(7) Express the present age of the Universe, t_0, in terms of the present Hubble constant, $H_0(= H(t_0))$. By using $H_0 = 72\,\mathrm{km/(s \cdot Mpc)}$, find the value of t_0 to two significant figures in the unit of year. Furthermore, by using $G = 6.67 \times 10^{-11}\mathrm{N \cdot m^2/kg^2}$, find the value of ρ_0 to two significant figures. Here, $1\,\mathrm{Mpc} = 10^6\,\mathrm{pc}$ and $1\,\mathrm{pc} = 3.09 \times 10^{16}\,\mathrm{m}$ (Mpc and pc are abbreviations of mega parsec and parsec, respectively), and one parsec is defined as the distance at which there is an angular separation of one arc-second $(= \frac{1}{3600}^\circ)$ between two objects $1.50 \times 10^{11}\,\mathrm{m}$ apart. The distance $1.50 \times 10^{11}\,\mathrm{m}$ is the average distance between the Earth and the Sun.

(the 2nd Challenge)

Solution

(1) When the atom (the source of light) recedes from the observer, the wavelength of the observed light, λ, becomes longer:

$$\lambda = \frac{c+v}{c}\lambda_0.$$

(2) We divide both sides of Eq. (2.74) by r_0 and use $\frac{1}{r_0}v(t) = \frac{1}{r_0}\frac{dr}{dt} = \frac{da}{dt}$ and $\frac{r}{r_0} = a$ to obtain

$$H(t) = \frac{1}{a}\frac{da}{dt}. \tag{2.77}$$

(3) The mass within the radius r is $\rho\frac{4}{3}\pi r^3$. It exerts a gravitational force on particle P. In terms of m, the mass of P, the equation of motion of P is

$$m\frac{d^2r}{dt^2} = -G\frac{\rho\frac{4}{3}\pi r^3 m}{r^2} = -\frac{4\pi Gm}{3}\rho r.$$

We divide both sides of the above by m and r and use Eq. (2.75) to obtain

$$\frac{1}{a}\frac{d^2a}{dt^2} = -\frac{4\pi G}{3}\rho.$$

(4) Particle P has a kinetic energy of $\frac{1}{2}mv^2 = \frac{1}{2}m(\frac{dr}{dt})^2$ and a gravitational potential energy of $-\frac{G \cdot \rho \frac{4\pi}{3} r^3 m}{r} = -\frac{4\pi G}{3}\rho r^2 m$. Then, by the law of conservation of mechanical energy, we have,

$$\frac{1}{2}m\left(\frac{dr}{dt}\right)^2 - \frac{4\pi G}{3}\rho r^2 m = 0$$

$$\therefore \left(\frac{dr}{dt}\right)^2 = \frac{8\pi G}{3}\rho r^2.$$

Next, we divide both sides of the preceding equation by r^2 and use Eq. (2.75) to obtain

$$\left(\frac{1}{a}\frac{da}{dt}\right)^2 = \frac{8\pi G}{3}\rho. \tag{2.78}$$

(5) Since the total mass is conserved we have

$$\rho \cdot \frac{4}{3}\pi r^3 = \rho_0 \cdot \frac{4}{3}\pi r_0^3$$

$$\therefore \rho = \frac{r_0^3}{r^3}\rho_0 = \frac{\rho_0}{a^3}. \tag{2.79}$$

(6) From Eqs. (2.78) and (2.79), we have

$$\left(\frac{da}{dt}\right)^2 = \frac{8\pi G}{3} \cdot \frac{\rho_0}{a}.$$

By using $a = (\frac{t}{t_0})^n$, we get

$$\frac{da}{dt} = \frac{k}{\sqrt{a}} = k\left(\frac{t}{t_0}\right)^{-\frac{n}{2}}. \tag{2.80}$$

Where $k = \sqrt{\frac{8\pi G}{3}\rho_0}$. Differentiating both sides of Eq. (2.76) with respect to t yields

$$\frac{da}{dt} = \frac{n}{t_0^n}t^{n-1}. \tag{2.81}$$

By comparing the exponent of t in the right hand side of Eq. (2.80) with that of Eq. (2.81), we obtain

$$n = \frac{2}{3}.$$

Then, Eq. (2.81) becomes $\frac{da}{dt} = \frac{2}{3t_0}(\frac{t}{t_0})^{-\frac{1}{3}} = \frac{2}{3t_0}\frac{1}{\sqrt{a}}$. So, we have

$$\frac{2}{3t_0} = k = \sqrt{\frac{8\pi G}{3}\rho_0}$$

$$\therefore t_0 = \frac{1}{\sqrt{6\pi\, G\rho_0}}. \tag{2.82}$$

(7) From $\frac{da}{dt} = \frac{2}{3t_0}(\frac{t}{t_0})^{-\frac{1}{3}}$, we have $(\frac{da}{dt})_{t=t_0} = \frac{2}{3t_0}$. Hence, by using Eq. (2.77) and $a(t_0) = 1$, we have

$$H_0 = H(t_0) = \left(\frac{1}{a}\frac{da}{dt}\right)_{t=t_0} = \frac{2}{3t_0}$$

$$\therefore t_0 = \frac{2}{3H_0}.$$

From $H_0 = \frac{7.2\times10^4}{3.09\times10^{16}\times10^6} \approx 2.33\times10^{-18}$ s^{-1} and 1 year $= 365 \times 24 \times 60 \times 60$ s $\approx 3.15 \times 10^7$ s, we get

$$t_0 = \frac{2}{3H_0} = \frac{2}{3 \times 2.33 \times 10^{-18}}\, \text{s} = 2.86 \times 10^{17}\, \text{s}$$

$$= 9.1 \times 10^9\ \text{year.}$$

Finally, from Eq. (2.82), we obtain

$$\rho_0 = \frac{1}{6\pi\, Gt_0^2}$$

$$= \frac{1}{6\pi \times 6.67 \times 10^{-11} \times (2.86 \times 10^{17})^2}\, \text{kg/m}^3$$

$$= 9.7 \times 10^{-27}\ \text{kg/m}^3.$$

Chapter 3

Oscillations and Waves

Elementary Course

3.1. Simple Harmonic Oscillation

When a particle of mass m at a displacement x from the origin, O, moves along the x-axis under the influence of a **restoring force** given by $-kx$ $(k > 0)$, the equation of motion of the particle is

$$m\frac{d^2x}{dt^2} = -kx. \tag{3.1}$$

By introducing arbitrary constants A and α, which are to be determined using initial conditions, we can write the solution of Eq. (3.1) as

$$x(t) = A\cos(\omega_0 t + \alpha), \tag{3.2}$$

where $\omega_0 = \sqrt{k/m}$.

Equation (3.2) indicates that the particle oscillates within a region $-A \leq x \leq A$. This motion is called a **simple harmonic oscillation**. Here, the constant A is called the **amplitude**, $\omega_0 t + \alpha$ is called the **phase**, the constant α is called the **initial phase**, ω_0 is called the **angular frequency**, $T = (2\pi/\omega_0)$ is called the **period** (the time required for one complete oscillation) and (the number of oscillations in one second) $f(= 1/T = \omega_0/2\pi)$ is called the **frequency** or the **eigenfrequency**.

When a weight of mass m suspended by a spring with a spring constant of k is set to oscillate, the angular frequency of the oscillation is given by $\omega_0 = \sqrt{k/m}$. A simple pendulum of length

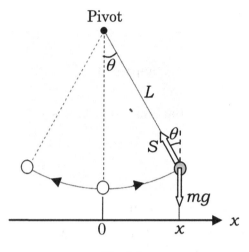

Fig. 3.1.

L exhibits a simple harmonic oscillation with an angular frequency of $\omega_0 = \sqrt{g/L}$, where g is the gravitational acceleration.

Example 3.1. Consider a simple pendulum of length L with a weight of mass m and let it swing with a small amplitude. Formulate the equation of motion for the oscillation of this system and derive its angular frequency.

Solution

As described in Fig. 3.1, we denote the displacement of the weight in the x-direction as x and the angle between the string and the vertical line as θ. The weight is under the action of a downward gravititional force of mg and the tension of the string toward the pivot. When θ is small, the weight moves very little vertically, so we can assume a static balance of force in the vertical direction:

$$S \cos \theta = mg. \tag{3.3}$$

The equation for the horizontal motion is

$$m\frac{d^2x}{dt^2} = -S \sin \theta. \tag{3.4}$$

When $|\theta| \ll 1$, we may apply the approximation $\cos\theta \approx 1$ to Eq. (3.3) to yield $S \approx mg$. From Fig. 3.1, we have $x = L\sin\theta$. Then, by substituting these relations into Eq. (3.4), we get

$$m\frac{d^2x}{dt^2} = -\frac{mg}{L}x. \tag{3.5}$$

This equation represents a simple harmonic oscillation so does Eq. (3.1), so by comparing Eq. (3.5) and Eq. (3.1), we get the angular frequency as

$$\omega_0 = \sqrt{\frac{g}{L}}. \tag{3.6}$$

This indicates that the angular frequency or the period of a simple pendulum is independent of the mass of the weight and the amplitude of oscillation. This is called Galileo's **isochronism of pendulum.** ∎

3.2. Waves

Wave is a phenomenon in which particles of a medium each oscillates about a point and the oscillation propagates to its neighboring particles of the medium in succession. When the oscillation can be represented by a sine function, the wave is specifically called a **sinusoidal wave.** When a particle simply oscillates about the origin, O, (defined as $x = 0$) with a period of T, then the displacement of the particle from O at time t, $y(0, t)$, can be written in the form of Eq. (3.2) as

$$y(0, t) = A\cos\left(2\pi\frac{t}{T}\right), \tag{3.7}$$

where, for simplicity, the initial phase is set to zero. Suppose this wave travels at a speed of V. Then, at a point P away from the origin by a distance x in the direction of the wave propagation, the medium is displaced by the same amount as that at the origin with a time delay of x/V. This means that the displacement at P at time t is equal to that at the origin at time $t - x/V$. Therefore the displacement of the medium at P can be expressed as

$$y(x,t) = A\cos\left\{2\pi\frac{1}{T}\left(t - \frac{x}{V}\right)\right\}. \tag{3.8}$$

The **wavelength**, λ, is the distance that the wave travels during a period, T, so we get the relation:

$$\lambda = VT \Leftrightarrow V = f\lambda \quad \text{(where } f = 1/T \text{ is the frequency)}. \tag{3.9}$$

Equation (3.8) can then be rewritten as

$$y(x,t) = A\cos\left\{2\pi\left(\frac{t}{T} - \frac{x}{\lambda}\right)\right\}. \tag{3.10}$$

We define the **wave number** k by the relation $k = 2\pi/\lambda$. Then, by using the relation $\omega = 2\pi/T$, we get

$$y(x,t) = A\cos(\omega t - kx). \tag{3.11}$$

Example 3.2.

(1) Find the amplitude, frequency, wavelength, and speed of propagation of the wave described by the equation

$$y = 0.2\cos\left[\pi(5t - 2x)\right].$$

Here, the units of length and time are taken to be meter and second, respectively.

(2) When a sinusoidal wave of amplitude 0.1 m and frequency 2 Hz travels at a speed of 2 m/s in the $-x$ direction, derive the expression for the displacement y at position x at time t by using an integer, n. Here, we assume that the displacement at the origin $(x = 0)$ at time $t = 0$ is zero, i.e., $y = 0$.

Solution

(1) From Eq. (3.10), the amplitude is $A = \underline{0.2\,\text{m}}$, the period is $T = \underline{0.4\,\text{s}}$, and the wavelength is $\lambda = \underline{1.0\,\text{m}}$. From these values, the frequency, $f = 1/T = \underline{2.5\,\text{Hz}}$, and the propagation speed of the wave, $V = f\lambda = \underline{2.5\,\text{m/s}}$.

(2) Using $A = 0.1\,\mathrm{m}$, $T = 0.5\,\mathrm{s}$ and $\lambda = VT = 1\,\mathrm{m}$ (because $V = 2\,\mathrm{m/s}$), we have

$$y(x,t) = 0.1\cos\left\{4\pi\left(t + \frac{x}{2}\right) + \left(n + \frac{1}{2}\right)\pi\right\}$$

$$= -0.1\sin\left\{4\pi\left(t + \frac{x}{2} + \frac{n\pi}{4}\right)\right\}. \qquad\blacksquare$$

Example 3.3. We observed a sinusoidal wave travelling along the x-axis at two positions ($x = 0$ and $x = 1\,\mathrm{m}$), and found the displacements of the wave at these two positions as follows:

$$y(0,t) = 0.2\cos(3\pi t), \quad y(1,t) = 0.2\cos(3\pi t + \pi/8).$$

Find the frequency, wavelength and speed of propagation of this wave in both cases in which the wave travels in the $+x$ and $-x$ directions, using the appropriate integer, n. Enumerate all the possible values of the wavelength longer than $0.7\,\mathrm{m}$.

Solution

A comparison of the observed data with Eq. (3.7) gives us the frequency, $f = \underline{1.5\,\mathrm{Hz}}$.

At the position $1\,\mathrm{m}$ away from the origin, the phase of the wave is shifted by $\frac{\pi}{8} \pm 2n\pi$ ($n = 0, 1, 2, \ldots$) from that at the origin (we can add an arbitrary integer multiple of 2π). If this phase difference is negative, its wave equation corresponds to Eq. (3.10) and can be regarded as a wave travelling in the $+x$ direction. In comparsion, if this phase difference is positive, it is a wave travelling in the $-x$ direction.

Therefore, in the case of the wave travelling in the $+x$ direction, we have from Eq. (3.10)

$$-\frac{2\pi}{\lambda} = \frac{\pi}{8} - 2n\pi, \quad (n = 1, 2, \ldots),$$

from which we get

$$\lambda = \frac{16}{16n - 1}\mathrm{m} = \underline{1.1\,\mathrm{m}}, \quad V = f\lambda = \frac{24}{16n - 1}\mathrm{m/s} = \underline{1.6\,\mathrm{m/s}},$$

where we substituted $n = 1$ from the condition that $\lambda > 0.7\,\mathrm{m}$.

In the case of the wave travelling in the $-x$ direction, we have from Eq. (3.10)

$$\frac{2\pi}{\lambda} = \frac{\pi}{8} + 2n\pi, \quad (n = 0, 1, 2, \ldots),$$

from which we get

$$\lambda = \frac{16}{16n + 1}\,\mathrm{m} = \underline{16\,\mathrm{m}}, \quad \underline{0.94\,\mathrm{m}},$$

$$V = f\lambda = \frac{24}{16n + 1}\,\mathrm{m/s} = \underline{24\,\mathrm{m/s}}, \underline{1.4\,\mathrm{m/s}},$$

where we substituted $n = 0$ and $n = 1$ from the condition that $\lambda > 0.7\,\mathrm{m}$. ∎

Elementary Problems

Problem 3.1.   A graph of a sinusoidal wave

Suppose we observe the wave propagation on a water surface. The displacement of the water surface at $t = 0$ is shown by the solid line in Fig. 3.2. After $0.2\,\mathrm{s}$ (i.e., at $t = 0.2\,\mathrm{s}$), a wave as shown by the dashed line in Fig. 3.2 is observed for the first time.

(1) What is the frequency of this wave? Choose one from the following. (a) $0.8\,\mathrm{Hz}$ (b) $1\,\mathrm{Hz}$ (c) $1.25\,\mathrm{Hz}$ (d) $1.5\,\mathrm{Hz}$ (e) $2.5\,\mathrm{Hz}$

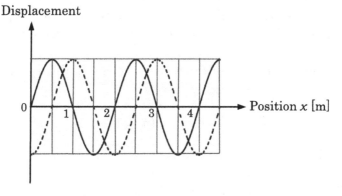

Fig. 3.2.

(2) Which of the following is the correct graph describing the change of the displacement at $x = 0$ with respect to time?

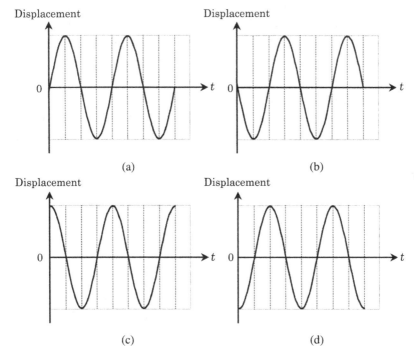

Fig. 3.3.

(the 1st Challenge)

Answer (1) (c), (2) (b)

Solution

(1) This wave travels a distance of a quarter of its wavelength in 0.2 s. Therefore, we find that the period, T, and the frequency, f, of this wave are

$$T = 0.2\,\text{s} \times 4 = 0.8\,\text{s}, f = 1/T = 1/0.8 = \underline{1.25\,\text{Hz}}.$$

(2) In Fig. 3.2, we see that the displacement at $x = 0$ is zero at $t = 0$, and it becomes negative just after that. Only (b) satisfies these conditions. ∎

Problem 3.2.   An observation of sound using microphones

A speaker and a microphone are placed as shown in Fig. 3.4. The speaker emits a sound of wavelength λ. The microphone detects the sound and converts it into an electric signal. In this way, we can obtain the waveform of the sound. We assume that the attenuation of the sound is negligible and that the microphone does not disturb the sound.

Curve (i) in Fig. 3.5 shows the waveform of the sound as measured by the microphone, which is set just in front of the speaker. We put another microphone at a distance $L = \lambda/4$ from the speaker and simultaneously measure the sound using the two microphones. Then, we observe both of the curves, (i) and (ii), shown in Fig. 3.5.

Speaker   **Microphone**

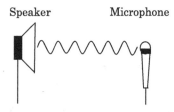

Fig. 3.4.

Output voltage (V)

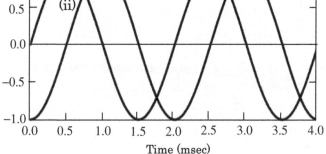

Time (msec)

Fig. 3.5.

(1) We put six microphones in series such that the distance between two neighboring microphones is L, as shown in Fig. 3.6. When $L = \lambda$, find the amplitude of the total combined output voltage from these six microphones.

(2) When $L = \frac{5}{6}\lambda$, find the combined output amplitude from microphones 1 and 4.

(3) When $L = \frac{5}{6}\lambda$, find the total combined output amplitude from all six microphones.

Speaker   Microphone 1   Microphone 2   Microphone 6

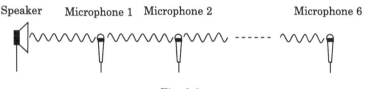

Fig. 3.6.

(4) Now, we array a large number of microphones in series in the same way as in part (1) such that the distance between two neighboring microphones is L. How does the total combined output amplitude change with the length L? From the graphs in Fig. 3.7, choose the best relation between the total combined output amplitude and the length L.

(the 1st Challenge)

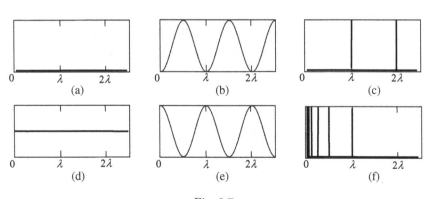

Fig. 3.7.

Answer (1) 6.0 V, (2) 0 V, (3) 0 V, (4) (c)

Solution

(1) From Eq. (3.10), we can see that these six outputs have the same phase. Thus, the total combined output amplitude is 6.0 V.

(2) The distance between microphones 1 and 4 is $3L = 3 \times \frac{5}{6}\lambda = \frac{5}{2}\lambda$. Using Eq. (3.10), we obtain that the phase difference is 5π. Therefore, their outputs cancel out each other, and the net output is 0 V.

(3) In the same way as part (2), it turns out that the combined output amplitude of microphones 2 and 5 is 0 V, and that of the microphones 3 and 6 is 0 V. Therefore, the total combined output amplitudes from all six microphones is 0 V.

(4) Unless $L =$ integer $\times \lambda$, the total output is zero, because for each microphone there exists another microphone whose output has the opposite phase. Only when $L =$ integer $\times \lambda$, all of the outputs have the same phase, and the combined output amplitude has a large non-zero value. The answer is (c). ∎

Advanced Course

3.3. Superposition of Waves

3.3.1. The Young's Double-Slit Experiment

Imagine a Young's double-slit experiment in vacuum, as shown in Fig. 3.8. The two light waves from slits S_1 and S_2 are superposed on the screen, and is observed at a point, P, on the screen. Let us discuss

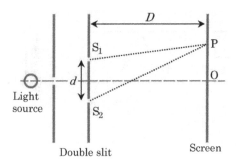

Fig. 3.8.

the intensity of the observed light. If we denote the oscillation of the light wave from slit S_1 at point P as

$$y_1 = A_1 \sin \omega t,$$

then that from slit S_2 at point P can be written as

$$y_2 = A_2 \sin(\omega t - \delta),$$

because the phase of y_2 is shifted by $\delta = 2\pi L/\lambda$, (where λ is the wavelength) due to the path difference $L = |S_2P - S_1P|$. Therefore, we obtain the expression for the resultant oscillation at P as

$$y = y_1 + y_2 = A_1 \sin \omega t + A_2 \sin(\omega t - \delta)$$
$$= \sqrt{A_1^2 + A_2^2 + 2A_1 A_2 \cos \delta} \cdot \sin(\omega t + \beta), \qquad (3.12)$$

where we have used trigonometric identities, and β is a constant dependent on δ.

The intensity of light is proportional to the square of its amplitude. Then, from Eq. (3.12), the intensity, I, of the observed light at P is

$$A_1^2 + A_2^2 + 2A_1 A_2 \cos \delta \propto I = I_1 + I_2 + 2\sqrt{I_1 I_2} \cdot \cos \delta, \qquad (3.13)$$

where I_1 and I_2 denote the intensities of the light waves y_1 and y_2, respectively. Let n be an integer, then from Eq. (3.13), we can see the following:

When $\delta = 2n\pi$,

the intensity I takes the maximum value of
$$I = (\sqrt{I_1} + \sqrt{I_2})^2 \quad \text{(bright fringes)} \qquad (3.14)$$
When $\delta = (2n + 1)\pi$,

the intensity I takes the minimum value of
$$I = (\sqrt{I_1} - \sqrt{I_2})^2 \quad \text{(dark fringes)}. \qquad (3.15)$$

This is the effect of the interference of the light waves. The situation (3.14) is called **constructive interference**, and (3.15) is called **destructive interference**. These interference conditions can be rewritten in terms of the path difference L as $L = n\lambda$ and $L = (n + 1/2)\lambda$, respectively.

Next, we consider the case where δ, the phase difference between y_1 and y_2, changes randomly with time. The average value of $\cos \delta$ over the observation time is zero, that is, $\cos \delta = 0$ in Eq. (3.13). Then, we obtain

$$I = I_1 + I_2.$$

In this case, the two waves do not interfere with each other, and the total intensitly is a simple sum of the intensities of the individual light waves. In order to interfere, the phase difference must be kept at a certain constant value. Therefore, two light waves from two independent sources, which imply their phase difference is randomly changing, do not cause any interference effect.

Example 3.4. In Fig. 3.8, let d be the interval between the two slits S_1 and S_2, and let D be the distance between the slits and the screen. Find the interval between two neighboring bright fringes observed on the screen, assuming $D \gg d$. You may use the following approximation formula:

$$(1 + x)^\alpha \approx 1 + \alpha x \text{ if } |x| \ll 1.$$

Solution

Let O be the point of intersection between the central axis and the screen and x be the length of OP. Then, the path difference, L, at P is

$$L = \sqrt{D^2 + \left(x + \frac{d}{2}\right)^2} - \sqrt{D^2 + \left(x - \frac{d}{2}\right)^2}$$

$$= D\left\{1 + \left(\frac{x + d/2}{D}\right)^2\right\}^{1/2} - D\left\{1 + \left(\frac{x - d/2}{D}\right)^2\right\}^{1/2}$$

$$\approx D\left\{1 + \frac{1}{2}\left(\frac{x + d/2}{D}\right)^2\right\} - D\left\{1 + \frac{1}{2}\left(\frac{x - d/2}{D}\right)^2\right\} = \frac{d}{D}x.$$

$$(3.16)$$

The constructive interference occurs at x where the path difference, L, satisfies the condition $L = n\lambda$, i.e., $x = \frac{n\lambda D}{d}$. Therefore, we obtain the interval between two neighboring bright fringes as

$$x_{n+1} - x_n = \frac{D}{d}\lambda. \tag{3.17}$$

■

3.3.2. *Standing Waves*

Let us consider the situation in which two waves are travelling in opposite directions with the same period, T, the same amplitude, A, and the same wavelength, λ, i.e., their velocities are also the same. We can write the oscillations of these waves as

the wave travelling in $+ x$ direction :

$$y_1(x,t) = A \sin\left\{\frac{2\pi}{T}\left(t - \frac{x}{v}\right)\right\}$$

the wave travelling in $- x$ direction :

$$y_2(x,t) = A \sin\left\{\frac{2\pi}{T}\left(t + \frac{x}{v}\right)\right\}.$$

These waves are superposed and interfere with each other. We can calculate the resultant wave $y(x,\,t)$ as

$$y(x,t) = y_1(x,t) + y_2(x,t) = \underbrace{2A\cos\left(\frac{2\pi}{\lambda}x\right)}_{\text{the amplitude term}} \cdot \underbrace{\sin\left(\frac{2\pi}{T}t\right)}_{\text{the oscillating term}},$$

$$\tag{3.18}$$

where we have used trigonometric identities as well as the relation $vT = \lambda$. When we observe the resultant wave at a point x, it oscillates according to $\sin(\frac{2\pi}{T}t)$, while its amplitude depends on x. Since this wave is oscillating in time and in space independently (the oscillation in time and that in space are independent of each other), it is not a traveling wave. Such a wave is called a **standing wave**, which, at every point, oscillates with time in phase. The locations

$$x = \left(n + \frac{1}{2}\right)\frac{\lambda}{2} \quad \text{for integers } n,$$

where the amplitude term that is equal to zero, are called **nodes**, and the locations

$$x = n\frac{\lambda}{2} \quad \text{for integers } n,$$

where the amplitude term that is at its maximium, are called **antinodes**. At an antinode, the amplitude is $2A$, which is the double of that of the original sinusoidal wave. We can also see that the distance between two adjacent nodes or antinodes is $\frac{\lambda}{2}$.

Example 3.5. Let us discuss a sinusoidal wave moving in the $+x$ direction.

$$y_1(x,t) = A \sin \frac{2\pi}{T}\left(t - \frac{x}{v}\right).$$

A wall located at $x = L$ reflects this wave. Answer the following questions for each of the following two cases: (a) The wall is a fixed end. (b) The wall is a free end.

(1) Find the expression for the reflected wave.
(2) Find the expression for the resultant wave produced by the incident wave and the reflected wave.

Note that a fixed end is an end where the amplitude of the resultant wave of the incident and reflected waves vanishes at all times, whereas a free end is an end where the displacement of the reflected wave is equal to that of the incident wave.

Solution

(1) The reflected wave moves in the $-x$ direction, and has the same amplitude, period, and velocity as the incident wave. Therefore, the reflected wave can be written as

$$y_2(x,t) = A \sin\left\{\frac{2\pi}{T}\left(t + \frac{x}{v} + \beta\right)\right\},$$

where β is a constant that is to be determined by the condition that the wave should satisfy at the end $x = L$ (a boundary condition).

(a) In the case where $x = L$ is a fixed end, the displacement y must always satisfy the condition $y(L, t) = y_1(L, t) + y_2(L, t) = 0$. Therefore,

$$\sin \frac{2\pi}{T} \left(t - \frac{L}{v} \right) + \sin \left\{ \frac{2\pi}{T} \left(t + \frac{L}{v} + \beta \right) \right\} = 0.$$

From this equation, β is determined to be $\beta = \frac{T}{2} - \frac{2L}{v}$, and we obtain

$$y_2(x, t) = A \sin \left\{ \frac{2\pi}{T} \left(t + \frac{x - 2L}{v} \right) + \pi \right\}$$

$$= -A \sin \left\{ \frac{2\pi}{T} \left(t + \frac{x - 2L}{v} \right) \right\}.$$

(b) On the other hand, in the case where $x = L$ is a free end, the displacement y must satisfy the condition. $y_1(L, t) = y_2(L, t)$. Therefore,

$$\sin \frac{2\pi}{T} \left(t - \frac{L}{v} \right) = \sin \left\{ \frac{2\pi}{T} \left(t + \frac{L}{v} + \beta \right) \right\}.$$

From this equation, β is determined to be $\beta = -\frac{2L}{v}$, and we obtain

$$y_2(x, t) = A \sin \left\{ \frac{2\pi}{T} \left(t + \frac{x - 2L}{v} \right) \right\}.$$

(2) The resultant wave is $y(x, t) = y_1(x, t) + y_2(x, t)$. Then, we have

(a) $$y(x, t) = 2A \sin \left(\frac{2\pi}{T} \cdot \frac{L - x}{v} \right) \cdot \cos \left\{ \frac{2\pi}{T} \left(t - \frac{L}{v} \right) \right\}$$

(b) $$y(x, t) = 2A \cos \left(\frac{2\pi}{T} \cdot \frac{L - x}{v} \right) \cdot \sin \left\{ \frac{2\pi}{T} \left(t - \frac{L}{v} \right) \right\}.$$

We can see that these resultant waves are standing waves, oscillating independently in time and in space. ∎

3.3.3. *Beats*

We now consider the superposition of two waves with slightly different frequencies. Suppose that the frequencies of these two waves are f_1 and f_2, respectively, and that the waves are expressed as $y_1(t) = A\sin(2\pi f_1 t)$ and $y_2(t) = A\sin(2\pi f_2 t + \alpha)$ at a point. Then, using the trigonometric identities, the superposed wave can be written as

$$y(t) = y_1(t) + y_2(t) = \underbrace{2A\cos\left(\frac{2\pi\,(f_1 - f_2)\,t - \alpha}{2}\right)}_{\text{the slowly varying amplitude term}}$$

$$\underbrace{\cdot\sin\left(\frac{2\pi\,(f_1 + f_2)\,t + \alpha}{2}\right)}_{\text{the oscillating term}}. \qquad (3.19)$$

While the sine function in Eq. (3.19) oscillates at a high frequency, the cosine function varies slowly with time since $|f_1 - f_2|$ is small. This phenomenon, illustrated in Fig. 3.9, is called **beats**. We do not hear two sounds at two different frequencies f_1 and f_2 separately, but hear the superposed wave whose frequency is the average of the original frequencies and whose amplitude is periodically and slowly varying. The period of the variation in amplitudes, T, which is the period of the beats, is the interval of time satisfying the

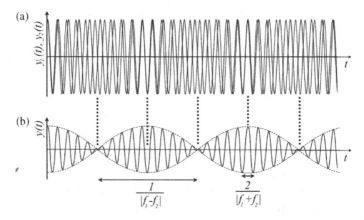

Fig. 3.9.

cosine function equal to zero. Therefore, the expressions for the beat period, T, and the beat frequency, f, are

$$T = \frac{1}{|f_1 - f_2|}, \quad f = \frac{1}{T} = |f_1 - f_2|, \tag{3.20}$$

respectively.

3.4. The Doppler Effect

When a source of sound is in motion relative to an observer, the observer hears the sound at a frequency different from the original one. This effect is called the **Doppler effect**. It occurs not only with sound but also with light.

Let us consider a source emitting a sound of frequency f, as shown in Fig. 3.10(a). Let V be the velocity of the sound, v_S be the velocity of the source, and suppose that $v_S < V$. Note that V is a constant regardless of the velocity of the source because the emitted sound propagates in an air that is at rest. In Fig. 3.10(a), wave crests are shown. We can see that the wavelength in front of a

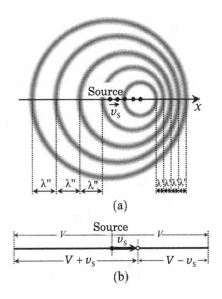

Fig. 3.10.

moving source, λ', is shorter and the wavelength behind the source, λ'', is longer ($\lambda' < \lambda < \lambda''$). As shown in Fig. 3.10 (b), the wave crests travel a distance, V, per unit time, whereas the source travels v_S. Since f wave crests are produced per unit time, there are in a unit time f wave crests in the interval $V - v_S$ in front of the source and in the interval $V + v_S$ behind the source. Therefore, the wavelengths λ' and λ'' are

$$\lambda' = \frac{V - v_S}{f} = \left(1 - \frac{v_S}{V}\right)\lambda, \; \lambda'' = \frac{V + v_S}{f} = \left(1 + \frac{v_S}{V}\right)\lambda, \quad (3.21)$$

respectively. Here $\lambda = V/f$ is the wavelength of the sound emitted from the source when it is at rest. Therefore, a stationary observer hears the sound wave with the following frequencies:

$$f' = \frac{V}{\lambda'} = \frac{V}{V - v_S}f \quad \text{(in front of the moving source)}, \quad (3.22)$$

$$f'' = \frac{V}{\lambda''} = \frac{V}{V + v_S}f \quad \text{(behind the moving source)}. \quad (3.23)$$

Next, let us discuss the case where both the source and the observer move. Suppose that the source moves to the right at a velocity, v_S, and that the observer also moves to the right at a velocity, v_0, in front of the moving source. The wavelength of the sound as heard by this observer is unchanged by his motion. It is the λ' given by the first equation of Eq. (3.21). However, the velocity of the sound will be replaced by the relative velocity $V - v_0$, due to the motion of the observer. Thus, in this case, we obtain the following expression for the frequency f' rather than Eq. (3.22):

$$f' = \frac{V - v_0}{\lambda'} = \frac{V - v_0}{V - v_S}f. \quad (3.24)$$

In the case of $v_S \ll V$ and $v_0 \ll V$, Eq. (3.24) is approximately written as

$$f' = \frac{1 - v_0/V}{1 - v_S/V}f \approx \left(1 - \frac{v_0}{V}\right)\left(1 + \frac{v_S}{V}\right)f \approx \left(1 - \frac{v_0 - v_S}{V}\right)f.$$
$$(3.25)$$

This result shows that the change in the observed frequency due to the motion of the observer is determined by his velocity relative to the source, $v_0 - v_S$.

3.4.1. *The Doppler Effect of Light*

In the case of light, the velocity of light relative to an observer must always be kept constant due to the principle of constancy of light velocity in the theory of special relativity. Thus, the wave velocity relative to the observer, $V - v_0$, in Eq. (3.24) should now be replaced by the constant light velocity, c. When the source of light is moving away from the observer at a velocity of v_S, we obtain

$$f' = \frac{c}{c + v_S} f. \tag{3.26}$$

Note that this result depends on the velocity of the source relative to the observer, but is independent of the velocity of the observer. In fact, Eq. (3.26) is an approximate expression (that is valid only in the case when $v_S \ll c$) for the general expression for f', which can be derived from the theory of special relativity as

$$f' = \sqrt{\frac{c - v_S}{c + v_S}} f. \tag{3.27}$$

We can see that in the case of $v_S \ll c$, Eqs. (3.26) and (3.27) are equal up to the first order with respect to v_S/c.

3.4.2. *Shock Waves*

If a source of wave moves faster than the wave velocity ($v_S > V$), the situation changes dramatically. As shown in Fig. 3.11, as the source moves, it generates spherical waves at each point. These spherical waves form a cone-shaped wave crest, where the spherical waves constructively interfere with one another, and the amplitude of the cone-shaped wave becomes extremely large. This pulse-like wave, advancing in the direction of the arrows in Fig. 3.11, is called a **shock wave**. While the source moves a distance $v_S t$ from S_0 to S during time t, the wave travels Vt from S_0 to A. Therefore, the half

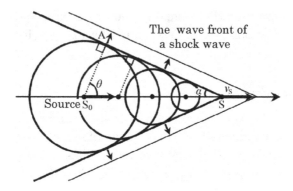

Fig. 3.11.

apex angle of the cone, α, is given by

$$\sin \alpha = \frac{V}{v_{\mathrm{S}}}.$$

In particular, when $v_{\mathrm{S}} = V$, we have $\alpha = 90°$, which means that in this case, the wave crest becomes perpendicular to the direction of the source movement. As illustrated in Fig. 3.11, the propagating direction of the shock wave makes an angle of θ to the direction of the velocity of the source, where θ is given by

$$\cos \theta = \frac{V}{v_{\mathrm{S}}}.$$

Advanced Problems

Problem 3.3.　The propagation velocity of a water wave

Wave motion is one of the broadest scientific subjects. The behavior of water waves and the propagation characteristics of various kinds of waves, such as light and sound, are fascinating, yet familiar in everyday experiences. Here, we take up a few attractive problems specific to water waves, which are, in general, associated with water motions in a gravitational field.

The dynamics of water waves is relatively complicated, compared to other waves. This complication results from the fact that the periodic motion of water is essentially two-dimensional. For comparison, let us look at a couple of typical examples of waves. Waves propagating on strings are transverse waves, in which the media of strings oscillate perpendicularly to the direction of wave propagation. On the other hand, sound waves propagating in the air are longitudinal waves, in which the medium of air oscillates parallel to the direction of wave propagation. However, it should be noted that motions in water waves are neither only perpendicular nor only parallel to the direction of wave propagation, but are perpendicular and parallel simultaneously.

Since the surface of water goes up and down, the motion of water undoubtedly has a component perpendicular to the direction of wave propagation. Next, consider a person floating with a swimming ring in the sea. When he is floating at a crest of the wave, he moves toward the seashore; in contrast when he is floating at a trough of the wave, he moves away from the seashore. So, we can see from this example that wave motion in the direction parallel to the wave propagation surely exists in water waves.

Although the water motion mentioned above is, in general, described by a motion along an elliptical orbit in a vertical plane. For simplicity, we may consider this motion a circular motion at a constant speed as shown in Fig. 3.12. Here, waves move to the right at a constant speed, V.

In Figs. 3.12(a)–(c), the center lines of the circles, which are denoted by broken horizontal lines, are fixed in space. Figure 3.12(a) shows the situation in which the trough of the wave goes into the position just under the center, O. Then, a cluster of water (denoted as a black dot on the water surface) located at the bottom of the circle rotates clockwise with time along a circular path, and then arrives at the height of the center O. The surrounding water also rises together with this cluster of water, as shown in Fig. 3.12(b). As time goes on, the cluster of water (denoted as a black dot) arrives at the position just above the center O, the position at the top of the wave surface, as shown in Fig. 3.12(c).

In the neighborhood of the surface of a water wave, a cluster of water at each position moves along a circular path, as shown in

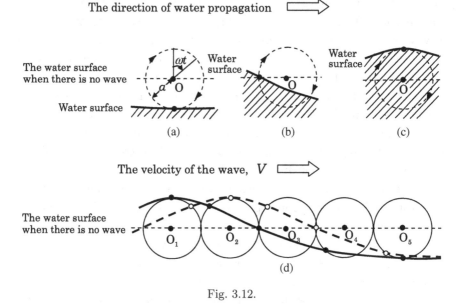

Fig. 3.12.

Fig. 3.12(d). Suppose the water surface at a certain moment is shown by the solid wavy line in Fig. 3.12(d). Clusters of water on the surface, some of which are shown by black dots, follow circular orbits and move to the positions shown by white dots. The broken wavy line formed by connecting these white dots indicates a new water surface. Note that the water surface represented by the broken wavy line is also obtained by translating the water surface represented by the solid wavy line. In the following problems, we denote the radius of the circular orbit as a and the angular frequency of the circular motion of water as ω (unit: radian).

(1) The period of a wave is that of the oscillating medium. We denote the wavelength as λ. Express the velocity of a wave, V, in terms of its amplitude a; its angular frequency, ω; or its wavelength, λ.

Let us look at the behavior of a water wave, particularly, its circular motion in the frame of reference moving to the right at a velocity, V, together with the propagation of the water wave, as shown in Fig. 3.13. In this frame of reference, the water surface

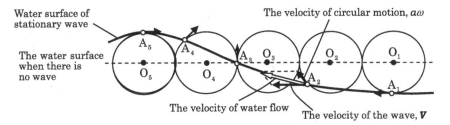

Fig. 3.13.

is at rest, keeping its waveform. In the coordinate used in Fig. 3.12, the centers of the circular motion of water clusters are fixed on a horizontal axis. However, we note that the centers of the circles in Fig. 3.13 move to the left at the velocity V. Therefore, the motion of water in this frame is described by the superposition of the circular motion of radius a and the horizontal motion to the left at the velocity V.

As shown in Fig. 3.13, there is an arrow on each white dot directed along the tangential line of its corresponding circle, and it denotes the velocity vector of the circular motion of the cluster of water at that point. The net velocity vector is the sum of two vectors, namely, the previously mentioned velocity vector of magnitude $a\omega$ (circular motion) and the uniform velocity moving toward the left V. Also, we should note that at each point, the direction of the sum of these two vectors corresponds to that of the water surface. As an example, the net velocity vector at the white small dot A_2 on the circle O_2 is the vector sum (shown by a white-arrow vector) of two black-arrow vectors in Fig. 3.13. So, it turns out that in this reference frame, clusters of water move up and down along the water surface. However, we note that this description is valid only when the magnitude of the velocity vector of the circular motion is smaller than the magnitude of the uniform velocity vector, namely, i.e., only when the relation $a\omega < V$ is satisfied.

(2) Suppose each white dot shown in Fig. 3.13 represents a cluster of water with mass Δm. When the clusters are located at the two positions A_1 and A_5, derive the expressions for the speeds at these two positions in terms of physical quantities a, ω

and V. Then, write down the equation that expresses the law of conservation of mechanical energy. Notice that no forces other than the gravitational force are acting on the water. Denote the gravitational acceleration as g.

(3) By using the results of part (2), derive the relation between the velocity, V, and the wavelength, λ.

(4) Using the result of part (3), evaluate the wavelengths of two waves with periods of $5\,\mathrm{s}$ and $10\,\mathrm{s}$, respectively. Take the gravitational acceleration as $g = 9.80\,\mathrm{m/s}^2$.

Generally speaking, the laws in physics are relations between various physical quantities. The units for mass, [M]; length, [L]; and time, [T], are the fundamental units in mechanics. Some derived units such as those for velocity, $[\mathrm{LT}^{-1}]$, acceleration, $[\mathrm{LT}^{-2}]$; density, $[\mathrm{ML}^{-3}]$; etc., are also used. Suppose quantity "C" has a unit that is the product of the pth power of the unit of length, the qth power of the unit of time and the rth power of the unit of mass, i.e., $[C] = [\mathrm{L}^p\mathrm{T}^q\mathrm{M}^r]$. Such a relation between units is called a dimensional relation. Analysis based on dimensional relations is frequently useful to derive relations between physical quantities.

(5) Assume that the wave velocity, V, depends only on the density of water, ρ; the gravitational acceleration, g; and the wavelength, λ. Then, the dimension of V is $[V] = [\rho^p][g^q][\lambda^r]$. By rewriting this relation in terms of the fundamental units, first find p, q and r, and then confirm the result of part (3). If a dimensionless number (constant) is involved in the result of part (3), regard it as unity and compare the two results.

(the 2nd challenge)

Solution

(1) We denote the wave frequency as f. Using the relations $V = f\lambda$ and $\omega = 2\pi f$, the wave velocity is

$$V = \frac{\lambda\omega}{2\pi}. \tag{3.28}$$

(2) Since the velocity at the crest A_5 of the wave is $V - a\omega$ and that at the trough A_1 is $V + a\omega$, from the law of conservation of mechanical energy, we have

$$\frac{1}{2}\Delta m \cdot (V - a\omega)^2 + 2\Delta m \cdot ga = \frac{1}{2}\Delta m \cdot (V + a\omega)^2. \quad (3.29)$$

(3) From Eq. (3.29), we have the relation $\omega V = g$. Using this relation together with Eq. (3.28), we obtain the relation between the wave velocity, V, and the wavelength, λ, as

$$V = \sqrt{\frac{g\lambda}{2\pi}}. \quad (3.30)$$

It turns out that the wave velocity does not depend on the depth of the water but depends only on the wavelength.

(4) From Eqs. (3.28) and (3.30), we obtain the relation $\lambda = \frac{g}{2\pi}T^2$. Therefore, the wavelengths are $\lambda = 39$ m for $T = 5$ s and $\lambda = 156$ m for $T = 10$ s, respectively.

(5) Since we assume that the wave velocity, V, depends only on the density, ρ; the gravitational acceleration, g; and the wavelength, λ, we consider the relation $V \propto \rho^p g^q \lambda^r$. If we represent the dimensions of time, length and mass by [T], [L] and [M], respectively, the dimensions of the velocity, the density, the gravitational acceleration and the wavelength are, respectively,

$$[V] = [L][T]^{-1}, \quad [\rho] = [M][L]^{-3}, \quad [g] = [L][T]^{-2}, \quad [\lambda] = [L].$$

Substituting these relations into the relation $V \propto \rho^p g^q \lambda^r$ yeilds

$$[T]^{-1}[L] = [T]^{-2q}[L]^{-3p+q+r}[M]^p.$$

From this relation, we obtain $p = 0, q = 1/2$ and $r = 1/2$. Finally, we have $V = \sqrt{g\lambda}$. This result agrees with that obtained in part (3), apart from a numerical factor. ∎

Problem 3.4.   The dispersion of light and refractive index

As shown in Fig. 3.14, a ray of white light incident on a triangular glass prism separates into red, yellow, and violet rays (in ascending

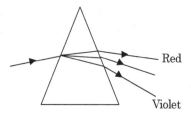

Fig. 3.14.

order of refraction angle). This is called the **dispersion of light**. It is caused by the different refractive indices of glass for different wavelengths of light, or equivalently, by the dependence of the speed of light in glass on its wavelength. Let us consider the cause of such phenomenon.

Light is an electromagnetic wave. The speed of light in a medium of permittivity ε and of permeability μ is, in general, given by $\frac{1}{\sqrt{\varepsilon\mu}}$. Let ε_0 and μ_0 be the vacuum permittivity and permeability, respectively, and let $\varepsilon_r = \frac{\varepsilon}{\varepsilon_0}$ be the relative permittivity of glass at a particular wavelength of lights λ. Let us assume that the relative permeability of glass $\frac{\mu}{\mu_0}$ is always unity, independent of the wavelength of light. If we denote the speed of light in vacuum as c, and that in a certain medium as c', then the refractive index of the medium, n, is defined as $n = \frac{c}{c'}$. Let us assume that $c = 2.998 \times 10^8 \text{m/s}$ and that the refractive index of air is equal to 1.

(1) Show that n_G, the refractive index of glass at wavelength λ, satisfies the relation

$$n_G^2 = \varepsilon_r.$$

(2) Let us consider a parallel-plate capacitor whose capacitance in vacuum is C_0. When the space between the plates is filled with a glass of relative permittivity ε_r, the capacitance of the capacitor becomes $C = \varepsilon_r C_0$. Let E_0 be the magnitude of the electric field between the plates of a capacitor of capacitance C_0 when an electric charge of Q is loaded onto the plates ($+Q$ on one plate, and $-Q$ on the other plate). Find E, the magnitude of the electric field between the plates when the parallel-plate capacitor

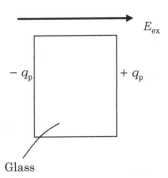

Fig. 3.15.

is filled with glass and electric charge of Q is loaded onto its plates. Express the answer in terms of E_0 and ε_r.

Here, we assume that the interval between the plates is sufficiently narrow as compared with their areas so that the electric field can be regarded as being perpendicular to the plates.

When an electric field, E_{ex}, is externally applied to a rectangular parallelepiped glass of relative permittivity ε_r, both positive and negative charges appear on the sides perpendicular to the electric field as in Fig. 3.15 (let q_p be the positive charge per unit area, and $-q_p$ be the negative one). This is comprehended as follows. The electric field causes a deviation of the centers of the positive and negative charges of the atoms composing the glass to yield the appearance of the charges on both sides. This is called **polarization**.

We assume that the polarization is caused by a displacement of a charged particle of mass m and charge $q(>0)$ from its equilibrium position $(x = 0)$. Further, we may assume that a restoring force of $-kx$, which is proportional to the displacement x, acts on the particle at the position x.

(3) Suppose the charged particle oscillates only under the restoring force given above. Denote the acceleration at the position x as α. Formulate the equation of motion of the particle and find the angular eigenfrequency, ω_0.

(4) When an oscillating electric field $E(t) = E_0 \cos \omega t \, (E_0 > 0)$ is applied to the charged particle considered in part (3), the particle

exhibits a simple harmonic oscillation and its position is given by $x(t) = A \cos \omega t$ where A is a constant. Express the amplitude of this oscillation, A, in terms of ω and ω_0.

(5) The polarization charges that appear on the sides of the glass by polarization, $\pm q_p$, are proportional to the displacement of the charged particles, x. Further the electric field in the glass produced by the polarization charge, E_p, is proportional to q_p. Thus, we can write $E_p = bx$ where b is a positive constant. Express n_G, the refractive index of glass, in terms of ω and ω_0. Furthermore, explain why the dispersion of light is caused by the glass prism.

(6) Let us assume that the model we considered can be applied to an optical glass called FK1. By using the measured refractive index of the glass for visible light in air, which is shown below, calculate ω_0, the angular eigenfrequency of the glass, and λ_0, the wavelength of light in air related to ω_0, to three significant figures.

Wavelength (m)	7.682×10^{-7}	5.876×10^{-7}	4.047×10^{-7}
Refractive Index	1.466	1.471	1.482

(the 1st Challenge)

Solution

(1) The speed of light in vacuum is $c = \frac{1}{\sqrt{\varepsilon_0 \mu_0}}$. Let the permittivity of glass at wavelength λ be ε and the permeability at wavelength λ be $\mu (= \mu_0)$. Since the speed of light with a wavelength of λ in glass is $c_G = \frac{1}{\sqrt{\varepsilon \mu}}$, we obtain

$$n_G = \frac{c}{c_G} = \sqrt{\frac{\varepsilon \mu}{\varepsilon_0 \mu_0}} = \sqrt{\varepsilon_r} \quad \therefore \quad n_G^2 = \varepsilon_r.$$

(2) The capacitance of the capacitor, C, is defined in terms of the voltage applied to the capacitor, V, and the stored charge, Q, as $C = \frac{Q}{V}$. Therefore, for a given stored charge, the voltage between

the plates is inversely proportional to the capacitance. Since the magnitude of the electric field is $E = \frac{V}{d}$ where d is the interval between the plates, we have, by denoting the voltage between the plates of the capacitor of capacitance C_0 as V_0,

$$\frac{E}{E_0} = \frac{V}{V_0} = \frac{C_0}{C} = \frac{C_0}{\varepsilon_r C_0} \quad \therefore \quad E = \frac{E_0}{\varepsilon_r}.$$

(3) The equation of motion of the charged particle at position x is

$$ma = -kx \quad \therefore \quad a = -\frac{k}{m}x = -\omega_0^2 x.$$

Therefore,

$$\omega_0 = \sqrt{\frac{k}{m}}.$$

(4) When the oscillating electric field is applied, the equation of motion of the charged particle is

$$ma = -kx + qE(t)$$
$$= -m\omega_0^2 x + qE_0 \cos \omega t.$$

Since $x(t) = A \cos \omega t$, the acceleration is

$$a = -A\omega^2 \cos \omega t.$$

Therefore, we have

$$A = \frac{qE_0}{m} \cdot \frac{1}{\omega_0^2 - \omega^2}.$$

(5) When an external electric field, $E_{\mathrm{ex}}(t)$, with an angular frequency of ω is applied to the glass, the electric field in the glass, $E(t)$, can be written by using the electric field caused by the polarization charge, $E_{\mathrm{p}}(t) = bx(t)$, as

$$E(t) = E_{\mathrm{ex}}(t) - E_{\mathrm{p}}(t).$$

Then, the relative permittivity of the glass, ε_r, becomes

$$\varepsilon_r = \frac{E_{ex}(t)}{E(t)} = 1 + \frac{E_p(t)}{E(t)} = 1 + b\frac{x(t)}{E(t)}$$

$$= 1 + b\frac{A}{E_0} = 1 + b\frac{q}{m} \cdot \frac{1}{\omega_0^2 - \omega^2}.$$

Hence, we get

$$n_G = \sqrt{\varepsilon_r} = \sqrt{1 + b\frac{q}{m} \cdot \frac{1}{\omega_0^2 - \omega^2}}. \tag{3.31}$$

The angular eigenfrequency of glass molecules, ω_0, is larger than that of visible light, ω. Therefore, the refractive index of glass for visible light, n_G, is larger than unity. Among visible light, the angular frequency of red light, which has a relatively long wavelength, is far away in value from the angular eigenfrequency of glass and this results in a relatively small refractive index, whence a small refractive angle. In contrast, the angular frequency of violet light, which has a relatively small wavelength, is close in value to the angular eigenfrequency of glass and this results in a relatively large refractive index, whence a large refractive angle.

(6) From Eq. (3.31),

$$n^2 = 1 + \frac{\omega_p^2}{\omega_0^2 - \omega^2}, \tag{3.32}$$

where $\omega_p^2 = b\frac{q}{m}$ and the angular frequency ω can be written in terms of wavelength, λ, as $\omega = \frac{2\pi c}{\lambda}$ (c is the speed of light in vacuum).

We denote the angular frequencies corresponding to some wavelengths λ_1 and λ_2 as ω_1 and ω_2, respectively. Then, from Eq. (3.32),

$$\omega_p^2 = (n_1^2 - 1)(\omega_0^2 - \omega_1^2)$$
$$= (n_2^2 - 1)(\omega_0^2 - \omega_2^2),$$

and we get

$$\omega_0^2 = \frac{(n_2^2 - 1)\omega_2^2 - (n_1^2 - 1)\omega_1^2}{n_2^2 - n_1^2}. \tag{3.33}$$

The wavelength of light in air related to ω_0, λ_0, is

$$\lambda_0 = \frac{2\pi c}{\omega_0}. \tag{3.34}$$

After substituting $\lambda_1 = 7.682 \times 10^{-7}$ m, $n_1 = 1.466$, $\lambda_2 = 5.876 \times 10^{-7}$ m, and $n_2 = 1.471$ in Eq. (3.33) and using Eq. (3.34), we get

$$\omega_{01} = \underline{1.86 \times 10^{16} \text{rad/s}}, \quad \lambda_{01} = \underline{1.01 \times 10^{-7} \text{ m}}.$$

Similarly, after substituting $\lambda_1 = 5.876 \times 10^{-7}$ m, $n_1 = 1.471$, $\lambda_2 = 4.047 \times 10^{-7}$ m, and $n_2 = 1.482$, we get

$$\omega_{02} = \underline{2.07 \times 10^{16} \text{ rad/s}}, \quad \lambda_{02} = \underline{0.91 \times 10^{-7} \text{ m}},$$

and finally, after substituting $\lambda_1 = 7.682 \times 10^{-7}$ m, $n_1 = 1.466$, $\lambda_2 = 4.047 \times 10^{-7}$ m, $n_2 = 1.482$, we get

$$\omega_{03} = \underline{2.01 \times 10^{16} \text{ rad/s}}, \quad \lambda_{03} = \underline{0.94 \times 10^{-7} \text{ m}}.$$

For visible light of wavelength $\lambda (> \lambda_0)$, the above simple model nearly explains the dispersion of light by a glass prism. ∎

Chapter 4

Electromagnetism

Elementary Course

4.1. Direct-Current Circuits

4.1.1. Electric Current and Resistance

Definition of the unit of current and Ohm's law

The unit of current is defined in terms of the attractive force between two currents flowing in a parallel direction (Fig. 4.1). This will be described in detail in Sec. 4.4.3 of the advanced course.

Suppose the attractive force acting between two identical straight, parallel currents located one meter apart from each other is 2×10^{-7} N for every meter of wire. Then, we define the amount of this current to be 1 A (ampere).

An electric current of 1 A carries an electric charge of 1 C (coulomb) in 1 s. When the energy needed to carry an electric charge of 1 C against a **potential difference** (also called **voltage**) is 1 J (joule), this potential difference is defined to have a value of 1 V (volt). When a voltage of 1 V is applied to a conductor, and a current of 1 A flows in the conductor, the electric resistance of the conductor

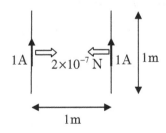

Fig. 4.1.

is defined to have a value of $1\,\Omega$ (ohm). The resistance of a conductor, R, is defined in terms of the voltage applied to the conductor, V, and the current flowing in the conductor, I, as $R = \frac{V}{I}$. This relation can alternatively be expressed as

$$V = RI. \tag{4.1}$$

When the resistance of the conductor, R, is independent of the applied voltage, V, or the electric current, I, it is said that the conductor satisfies **Ohm's law**, and such a resistance is called a **linear resistance** or an **ohmic resistance**. The resistance of some conductors varies with voltage or current, and such a resistance is called a **non-linear resistance** or a **non-ohmic resistance**.

Resistivity

The resistance of a conductor, R, is proportional to its length, l, and inversely proportional to its cross-sectional area, S:

$$R = \rho \frac{l}{S}. \tag{4.2}$$

Here, the factor ρ (its SI unit is $\Omega \cdot m$) is called the **resistivity** of the conductor. The value of resistivity depends on the material and its temperature.

4.1.2. *Resistors in Series and in Parallel*

When n resistors of resistances R_1, R_2, ..., R_n are connected to one another in **series**, the combined resistance, R, is equal to the sum of their individual resistances:

$$. \quad R = R_1 + R_2 + \cdots + R_n. \tag{4.3}$$

Resistance is a measure of the **difficulty for current flow** in the conductor.

When n resistors of resistances R_1, R_2, ..., R_n are connected to one another in **parallel**, the combined resistance, R, is given by the relation

$$\frac{1}{R} = \frac{1}{R_1} + \frac{1}{R_2} + \cdots + \frac{1}{R_n}, \tag{4.4}$$

where $\frac{1}{R}$ is called a **conductance**, a measure of the **ease for current flow** in the conductor. The unit of conductance is S (siemens). The total conductance of resistors connected in parallel is equal to the sum of their individual conductances.

Example 4.1. By using the definition of resistance, find the combined resistances of resistors R_1, R_2 connected to each other in series and in parallel.

Solution

In the case of series connection, as shown in Fig. 4.2, let the voltages applied to the resistances R_1 and R_2 be V_1 and V_2, respectively. The same amount of current, I, flows through R_1 and R_2. From the definition of resistance, Eq. (4.1), we get

$$V_1 = R_1 I, \quad V_2 = R_2 I.$$

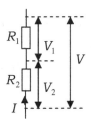

Fig. 4.2.

Since the total voltage, V, is $V = V_1 + V_2$, the combined resistance, R_s, is

$$R_s = \frac{V}{I} = \frac{V_1 + V_2}{I} = R_1 + R_2.$$

In the case of parallel connection, as shown in Fig. 4.3, the same voltage, V, is applied to R_1 and R_2. Let the currents flowing through R_1 and R_2 be I_1 and I_2, respectively. Then, we set $I_1 = \frac{V}{R_1}$ and $I_2 = \frac{V}{R_2}$. Let the combined resistance be R_p. Since the total current, I, is $I = I_1 + I_2$, from the relation $I = \frac{V}{R_p}$, we get

$$\frac{V}{R_p} = \frac{V}{R_1} + \frac{V}{R_2} \Rightarrow \frac{1}{R_p} = \frac{1}{R_1} + \frac{1}{R_2}.$$

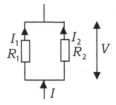

Fig. 4.3.

Then,

$$R_\text{p} = \frac{R_1 R_2}{R_1 + R_2}.$$ ∎

4.1.3. *Kirchhoff's Rules*

Kirchhoff's junction rule

At any junction in an electric circuit,

(The sum of incoming currents)
= (The sum of outgoing currents)   (4.5)

This relation is called Kirchhoff's junction rule. It implies the conservation of currents.

Kirchhoff's loop rule

The influence that makes a current flow from a lower to a higher potential is called **an electromotive force (emf)**.

In any closed loop of an electric circuit,

(The sum of emfs) = (The sum of voltage drops)   (4.6)

This relation is called Kirchhoff's loop rule. It implies that, once we select a point at which the electric potential–potential energy per unit charge — is set as zero, the electric potentials of any other points are automatically determined.

4.2. Magnetic Field and Electromagnetic Induction

Here, we will describe magnetic field and electromagnetic induction without using equations. Details are described in Sec. 4.4 of the advanced course.

4.2.1. *Magnetic Field*

A permanent magnet produces a special field around the magnet in which magnetic forces act. This field is called a **magnetic field** whose direction is the N-pole direction of a compass. An electric current produces a magnetic field that points clockwise around the current as shown in Fig. 4.4. If a current flows through a solenoid as shown in Fig. 4.5, a magnetic field that passes through the solenoid and points in the rightward direction is produced. The curves drawn to represent a magnetic field are called **magnetic field lines**. Magnetic field lines point outward from the N-pole of a permanent magnet and point inward to the S-pole; they neither disappear nor intersect one another (Fig. 4.6).

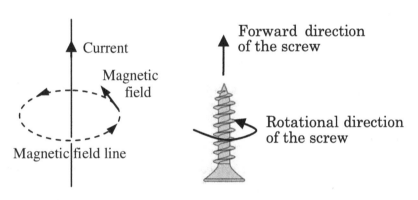

Fig. 4.4.

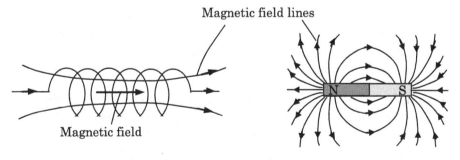

Fig. 4.5.

Fig. 4.6.

4.2.2. *Magnetic Force on Current*

A magnetic field exerts a force on a current. If the magnetic field is in the direction of the index finger of a left hand and the current is in the direction of the middle finger as shown in Fig 4.7, then the force on the current is in the direction of the thumb. This rule is called **Fleming's left-hand rule**. A motor utilizes the force on a current-carrying coil in a magnetic field to rotate the coil.

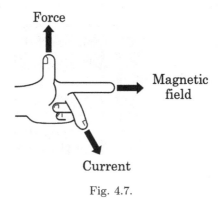

Fig. 4.7.

4.2.3. *Electromagnetic Induction*

As shown in Fig. 4.8, an emf appears in a coil when a bar magnet moves either toward or away from the coil. Then, an electric current flows in the direction of the emf. This emf is called an **induced emf**. This phenomenon is called **electromagnetic induction**. Electromagnetic induction appears when the number of magnetic field lines through the coil vary with time. The emf is induced to prevent the change in the magnetic field lines through the coil, and its magnitude is proportional to the instantaneous rate of change of the magnetic field lines through the coil.

Example 4.2. Which is the current direction, a or b, when the N pole of a bar magnet approaches a coil from the right side, as shown in Fig. 4.8? Which is the current direction, a or b, when the N pole of a bar magnet recedes from a coil to the right side?

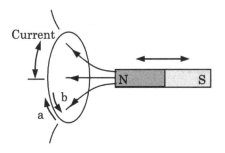

Fig. 4.8.

Solution

When the N pole of the bar magnet approaches the coil from the right side, the number of magnetic field lines through the coil pointing in the leftward direction increases. Then, a current in the direction of b̲ is induced so as to prevent the magnetic field lines from increasing. When the N pole of the bar magnet recedes from the coil to the right side, the number of magnetic field lines through the coil pointing in the leftward direction decreases. Then, a current in the direction of a̲ is induced so as to prevent the magnetic field lines from decreasing. ∎

Elementary Problems

Problem 4.1. A circuit with two batteries

Fill the boxes $\boxed{a}$ through $\boxed{g}$ in the following sentences with the appropriate numerical values. Suppose we make a circuit with two batteries (E_1, E_2) of emf $3\,\text{V}$ and with two resistors R_1, R_2 of resistance $6\,\Omega$ as shown in Fig. 4.9.

The electric potential at point C is $\boxed{a}$ V higher than that at point B. The electric potential at point A is $\boxed{b}$ V higher than that at point C. Therefore, the electric potential at point A is $\boxed{c}$ V higher than that at point B. Then, the current through R_2 is $\boxed{d}$ A and that through R_1 is $\boxed{e}$ A. Consequently, the current through E_2 is $\boxed{f}$ A and that through E_1 is $\boxed{g}$ A.

(the 1st Challenge)

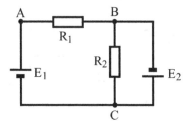

Fig. 4.9.

Answer a $= 3$, b $= 3$, c $= 6$, d $= 0.5$, e $= 1$, f $= 1.5$, g $= 1$

Solution

As shown in Fig. 4.10, let the current through R_1 flowing from point A to point B be I_1 and that through R_2 flowing from point C to point B be I_2. Also, let the potentials at points A, B and C be V_A, V_B and V_C, respectively.

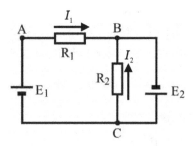

Fig. 4.10.

a. The potential at point C is $V_C - V_B = E_2 = \underline{3}\,\text{V}$ higher than that at point B.

b. The potential at point A is $V_A - V_C = E_1 = \underline{3}\,\text{V}$ higher than that at point C.

c. The potential at point A is $V_A - V_B = E_1 + E_2 = \underline{6}\,\text{V}$ higher than that at point B.

d. The current through R_2 is $I_2 = (V_C - V_B)/R_2 = \underline{0.5}\,\text{A}$.

e. The current through R_1 is $I_1 = (V_A - V_B)/R_1 = \underline{1}\,\text{A}$.

f. The current through E_2 is $I_1 + I_2 = \underline{1.5}\,\text{A}$.

g. The current through E_1 is $I_1 = \underline{1}\,\text{A}$.

■

Supplement

By using Kirchhoff's loop rule to the closed loop $E_1 \to A \to B \to C \to E_1$ and $E_2 \to C \to B \to E_2$, we get $E_1 = R_1 I_1 - R_2 I_2$ and $E_2 = R_2 I_2$. Substituting $E_1 = E_2 = 3\,\text{V}$ and $R_1 = R_2 = 6\Omega$ into the above equations, we get $I_1 = 1\,\text{A}$ and $I_2 = 0.5\,\text{A}$.

Problem 4.2.   A three-dimensional connection of resistors

Suppose we make a frame of a regular tetrahedron using six resistors with the same resistance, r, and apply a voltage of V between point O and point M as shown in Fig. 4.11. Then, a current, I, flows into point O and flows out of point M. Note that the resistance is proportional to the length of the resistor, and the contact resistance may be assumed to be negligibly small. M is the midpoint of resistor BC.

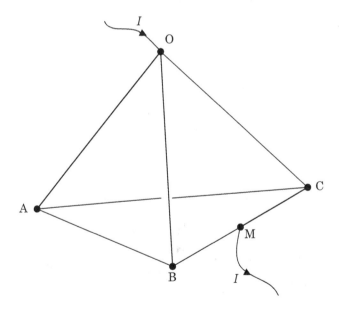

Fig. 4.11.

(1) Express the current that flows from point O to point B, i, in terms of the current I. In addition, express the currents from O

to C, from O to A, from A to B, from A to C, from B to M and from C to M in terms of I, respectively.

(2) Find the combined resistance between O and M.

(the 2nd Challenge)

Solution

(1) From the symmetry of the frame, we can assign the currents in the edge of the tetrahedron as shown in Fig. 4.12. By using Kirchhoff's loop rule, we get

$$ri = r\left(\frac{I}{2} - i\right) + r(I - 2i) = r\left(\frac{3}{2}I - 3i\right),$$

$$\therefore \ i = \frac{3}{8}I.$$

The currents are, respectively,

$$O \rightarrow C: \ i = \frac{3}{8}I, \quad O \rightarrow A: \ I - 2i = \frac{1}{4}I,$$

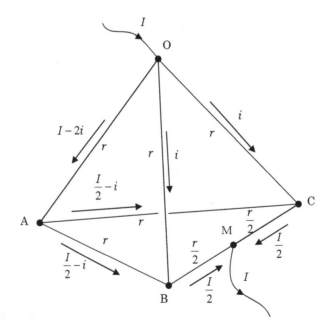

Fig. 4.12.

$$A \to B: \quad \frac{I}{2} - i = \underline{\frac{1}{8}I}, \quad A \to C: \quad \frac{I}{2} - i = \underline{\frac{1}{8}I},$$

$$B \to M: \quad \frac{I}{2}, \qquad\qquad C \to M: \quad \frac{I}{2}.$$

(2) The voltage between points O and M, V, is expressed as

$$V = \frac{r}{2} \cdot \frac{I}{2} + ri = \frac{rI}{4} + r \cdot \frac{3}{8}I = \frac{5}{8}rI,$$

$$\therefore \ R = \frac{V}{I} = \frac{\frac{5}{8}rI}{I} = \underline{\frac{5}{8}r}. \qquad\qquad \blacksquare$$

Alternative solution

From the symmetry of the frame, we can redraw the circuit as shown in Fig. 4.13, because the electric potential at point B is the same as

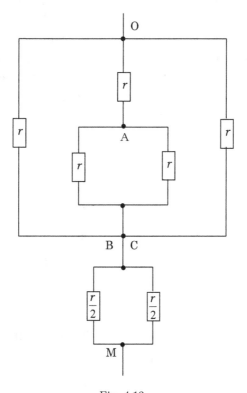

Fig. 4.13.

that at point C. Using the combined resistance in parallel connection, we find that the resistance between points A and B or C is equal to $r/2$. Then, we obtain the combined resistance between points O and B or C, R_1, as follows:

$$\frac{1}{R_1} = \frac{1}{r} + \frac{1}{r + \frac{r}{2}} + \frac{1}{r} \quad \therefore R_1 = \frac{3}{8}r.$$

The combined resistance between points B or C and M, R_2, is $R_2 = \frac{r}{4}$. Then, the total resistance is

$$R = R_1 + R_2 = \underline{\frac{5}{8}r}$$

Problem 4.3. A hand dynamo

A hand dynamo, shown in Fig. 4.14, can generate electricity when its handle is rotated, and a lamp can be lighted when connected to the terminal clips of a electricity-generating hand dynamo.

When we connect a battery to the clips of a dynamo, its handle rotates spontaneously. When we connect two hand dynamos by their clips to each other and rotate the handle of one dynamo, the handle of the other dynamo also rotates.

Terminal clips

Fig. 4.14.

(1) Which equipment has properties similar to a hand dynamo? Choose the best answer from (a) through (d).

(a) Battery
(b) Capacitor
(c) Motor
(d) Coil

When we connect a capacitor to the terminal clips of a hand dynamo and rotate its handle, the generated electricity is stored in the capacitor. After a sufficient amount of electricity is stored in the capacitor by several rotations of the handle, we release the handle.

(2) What will occur after the release? Choose the best answer from (a) through (d).

(a) The handle keeps rotating in the same direction as its previous rotation.
(b) The handle rotates in the direction counter to its previous rotation.
(c) The handle stops immediately.
(d) The handle rotates alternately between the same direction and the counter direction.

(3) Suppose we connect a hand dynamo, a lamp, a switch and a battery as shown in Fig. 4.15. When we switch on the circuit, the handle of the hand dynamo rotates and the lamp lightens. When we stop the handle, the lamp becomes brighter.
Choose the best reason from (a) through (e).

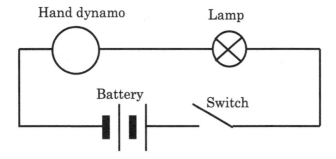

Fig. 4.15.

(a) The energy consumed to rotate the handle is utilized for lightening the lamp when the rotation is stopped.

(b) A large current is supplied to the circuit to rotate the stopped handle.

(c) Frictional heat energy that is generated by the rotation is used to lighten the lamp when the rotation is stopped.

(d) The internal resistance of the dynamo decreases when the handle is stopped, and a large current flows.

(e) The emf (in the counter direction) induced in the dynamo disappears when the handle is stopped, and a large current flows.

(the 1st Challenge)

Answer (1) (c), (2) (a), (3) (e)

Solution

(1) A hand dynamo has a motor inside; it is a motor with a handle connected to a coil. Permanent magnets are fixed inside the motor and by applying an external force on the handle, the coil rotates between the N and S poles of the magnets. The rotation of the coil changes the number of magnetic field lines through the coil, and an emf is induced in the coil. When we supply a current to the motor, the coil in the motor generates a magnetic field, and attractive and repulsive forces act between the coil and the permanent magnets. As a result, the coil rotates. When two hand dynamos are connected to each other by their terminal clips and one of the handle is rotated to generate electricity, an induced current flows into the motor of the other dynamo and rotates the handle that is connected to the coil of the other dynamo.

(2) After the capacitor is charged up by the electricity generated by the hand dynamo and the handle is released, a current flows from the capacitor into the motor of the dynamo. As a result, the handle continues to rotate until the capacitor discharges completely, and the current decays to zero. This current rotates the motor in the same direction as before. The rotation stops once the capacitor is completely discharged.

(3) When the rotation of the handle in the dynamo is stopped, the induced emf in the direction that reduces the induced current in the dynamo disappears. As a result, the total current in the circuit increases. Then, the work done by the battery increases (and the Joule heat generated in the dynamo also increases). ■

Advanced Course

4.3. Electric Charge and Electric Field

We say that there is an **electric field** in the space around a static **charged body** when there is an electric force acting on the body. We also say that a point **charge** produces an electric field around it because a point charge exerts electric forces on other charged bodies around it. When an electric force, F, is exerted on a test charge, q, we define the electric field (at the point where the charge is placed), E, by

$$F = qE. \tag{4.7}$$

4.3.1. *Gauss's Law*

We consider the electric field produced by a **point charge**, whose size is negligible, in vacuum. When a point test charge, q, is placed at P, a point located at a distance r from a point source charge, Q, the electric force that two charges q and Q exert on each other is

$$F = \frac{qQ}{4\pi\varepsilon_0 r^2}, \tag{4.8}$$

where ε_0 (epsilon zero) is a constant and is called the **permittivity of vacuum**. Equation (4.8) is called **Coulomb's law**. As shown in Fig. 4.16(a), the force between q and Q is repulsive when $qQ > 0$ and attractive when $qQ < 0$.

After dividing both sides of Eq. (4.8) by q, we get E, the electric field at point P:

$$E = \frac{Q}{4\pi\varepsilon_0 r^2}. \tag{4.9}$$

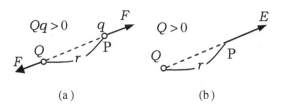

Fig. 4.16.

The electric field E points from the charge Q to point P when $Q > 0$ (see Fig. 4.16(b)).

It is useful to use **electric field lines** for visualizing an electric field. An electric field line is a curve or line whose tangent at point gives the direction of the electric field vector, $\boldsymbol{E}$, at that point. The number of electric field lines through a unit area perpendicular to $\boldsymbol{E}$ is equal to the magnitude of $\boldsymbol{E}$. Electric field lines are directed away from positive to negative charges, never intersect one another, and are never created nor annihilated in vacuum.

We call a flux of electric field lines an **electric flux**. The electric flux is a measure of the flow of an electric field and is expressed in terms of the number of electric field lines.

Example 4.3. An electric flux radiates out through a spherical surface of radius r centered on a positive point charge, Q. Show that the electric flux through the spherical surface is Q/ε_0.

Solution

As the magnitude of the electric field on the surface of a sphere of radius r is $\frac{Q}{4\pi\varepsilon_0 r^2}$, the electric flux through a unit area of the surface is $\frac{Q}{4\pi\varepsilon_0 r^2}$. The area of the sphere is $4\pi r^2$, so the total electric flux through the surface is

$$4\pi r^2 \times \frac{Q}{4\pi\varepsilon_0 r^2} = \frac{Q}{\varepsilon_0}. \tag{4.10}$$

■

This result is called **Gauss's law**. In the above, we have considered a spherical surface in which a point charge is at the center of the sphere, but we can obtain the same result even when the point charge is not at the center as long as it is inside the sphere. Further, it is not necessary that the shape of the closed surface be spherical. We can formulate Gauss's law in an integral form by integrating the inner product of E, the electric field vector, and ΔS, the vector perpendicular to a small surface ΔS (see Fig. 4.17):

$$\sum_{\Delta S} E \cdot \Delta S \Rightarrow \int_S E \cdot dS = \frac{Q}{\varepsilon_0}. \tag{4.11}$$

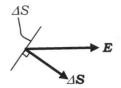

Fig. 4.17.

The integral on the left hand side of Eq. (4.11) is called a **surface integral**, and this result can be extended to any closed surface enclosing any number of charges. When $Q > 0$, the net electric flux is pointing outward from Q and is equal to Q/ε_0; (when $Q < 0$, the net electric flux is pointing inward toward Q and is equal to Q/ε_0). The value of the left-side integral depends on the total enclosed charge, which can either be distributed discretely or continuously.

Example 4.4. A positive electric charge is distributed uniformly along an infinitely long, thin and straight wire. The charge per unit length is $\lambda > 0$. As the wire is infinitely long, the generated electric field, E, is perpendicular to and symmetrically pointing outward from the wire. As shown in Fig. 4.18, we consider a cylindrical region of length L and radius r.

(1) Find the net value of the electric flux outward through the curved surface of the cylinder in terms of the magnitude of the electric field E on the surface, Further, find the total magnitude of the charge inside the cylinder in terms of λ.

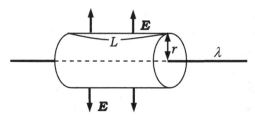

Fig. 4.18.

(2) Using Gauss's law, find the magnitude of the electric field E at a distance r from the wire.

Solution

(1) The electric field E is outward and uniform on the curved surface of the cylinder due to symmetry. Since the electric flux per unit area on the surface is E, the net value of the outward electric flux through the surface is equal to E multiplied by the surface area $2\pi r L$, and thus, the answer is $\underline{2\pi r L E}$.

The total charge contained in the cylinder is $\underline{\lambda L}$.

(2) By applying Gauss's law to the cylinder, we have

$$2\pi r L E = \lambda L / \varepsilon_0.$$

Thus, the magnitude of the electric field on the curved surface of the cylinder, E, is

$$E = \frac{\lambda}{2\pi \varepsilon_0 r}.$$ ∎

Example 4.5. Extend the previous example, Example 4.4, to a thin, flat infinite plane with a uniform surface charge per unit area, σ, as shown in Fig. 4.19. The electric field $\boldsymbol{E}$ is normal to the plane due to the infinite size of the plane, and it is symmetric on both sides of the plane. Find the magnitude of $\boldsymbol{E}$ at a distance r from the plane.

Solution

As shown in Fig. 4.19, we consider a cylinder (perpendicular to the plane) of a unit cross-sectional area. The electric field is normal to the

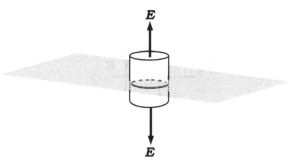

Fig. 4.19.

plane, so there is no flux through the curved surface of the cylinder. There is only an outward flux through both of the flat surfaces of the cylinder. Since we can write Gauss's law as $2E = \sigma/\varepsilon_0$, we have

$$E = \frac{\sigma}{2\varepsilon_0}. \tag{4.12}$$

Note that this electric field is independent of r. ∎

4.3.2. *Capacitors and Energy of Electric Field*

A **capacitor** is a device that stores electric charge. Metallic plates storing charge are called **capacitor plates**.

Example 4.6. As shown in Fig. 4.20, two charges of equal magnitude and opposite signs, $\pm Q$, are separately loaded onto two large parallel plates of area S. The distance between the plates is D.

(1) Find E, the magnitude of the electric field between the plates.
(2) Express U, the electrostatic energy stored in the capacitor, in terms of ε_0, E, D and S.
(3) The electrostatic energy U is considered the energy of the electric field. Find u, the energy of the electric field per unit volume between the two plates.
     We call this the **energy density** of the electric field.

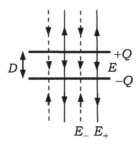

Fig. 4.20.

Solution

(1) The electric field outside the two parallel plates is zero because the two electric fields E_+ and E_- are of equal magnitude, $\sigma/2\varepsilon_0$, (since they are generated by the same charge density, $\sigma = Q/S$, on the plates) and cancel out (since they point in opposite directions). The electric field between the two parallel plates is given by the superposition of E_+ and E_-, which are in the same direction, and thereby is

$$E = E_+ + E_- = \frac{Q}{\varepsilon_0 S}. \tag{4.13}$$

(2) When the capacitor stores a charge of q, the electric field between the capacitor plates, E, is $E = q/\varepsilon_0 S$. In order to increase the stored charge from q to $q + dq$, it is necessary to move a small charge of dq against an electric force of Edq over a distance of D. Then, a work done of $EDdq$ is added to the energy of the capacitor. The total work needed to increase the charge of the capacitor from 0 to Q is

$$U = \int_0^Q EDdq = \frac{D}{\varepsilon_0 S} \int_0^Q qdq = \frac{D}{\varepsilon_0 S} \frac{Q^2}{2}.$$

This is the electrostatic energy stored in the capacitor.
By using $Q = \varepsilon_0 SE$, we have

$$U = SD \cdot \frac{1}{2}\varepsilon_0 E^2. \tag{4.14}$$

(3) As the volume of the capacitor is DS, the energy density of the electric field, u, is

$$u = \frac{1}{2}\varepsilon_0 E^2. \tag{4.15}$$

■

Suppose a charge of Q is stored in a capacitor to which an electric potential difference of V is applied. Then,

$$C = \frac{Q}{V}, \tag{4.16}$$

is called the **capacitance** of the capacitor. When an uniform electric field of magnitude E is produced between two parallel, charged plates separated by a distance of D, the potential difference between the plates, V, is $V = ED$. So, from Eqs. (4.13) and (4.16), we have

$$C = \frac{\varepsilon_0 S}{D}. \tag{4.17}$$

Further, the electrostatic energy of the capacitor is

$$U = \frac{1}{2}CV^2 = \frac{Q^2}{2C} = \frac{1}{2}QV. \tag{4.18}$$

4.4. Current and Magnetic Field

4.4.1. *Magnetic Field Generated by Current in Straight Wire*

We say that there is a **magnetic field** in the space around a moving charged particle when a magnetic force acts on the particle. The characteristic physical quantity for the magnetic field is the magnetic-flux-density vector, B. Hence, we use the quantity B as the magnetic-field vector. There are magnetic field lines, whose tangent gives the direction of the magnetic field, just like electric field lines. We define the magnitude of B as the number of magnetic field lines passing through a unit area perpendicular to B.

When a current of I flows in a straight conducting wire that is infinitely long, the magnitude of the magnetic field, B, at a distance

r from the wire in vacuum is

$$B = \frac{\mu_0 I}{2\pi r},\qquad(4.19)$$

where μ_0 is called the **permeability of vacuum**, and the magnetic field encircles the current in the clockwise direction (to which a right-handed screw that advances to the direction of the current rotates), as shown in Fig. 4.4. The speed of light in vacuum, c, is related to μ_0 and ε_0 as

$$c^2 = \frac{1}{\varepsilon_0 \mu_0}.\qquad(4.20)$$

(see Problem 4.4).

4.4.2. *Ampere's Law*

We consider a straight conductor carrying a current of I. The magnitude of the magnetic field, B, on the circumference of radius r in a plane perpendicular to the conductor can be written, from Eq. (4.19), as

$$B \cdot 2\pi r = \mu_0 I.\qquad(4.21)$$

That is, the product of the length of the circumference and B is equal to $\mu_0 I$. We can extend this relation to the following representation.

Let us consider a long straight current, I, and the magnetic field caused by it on a closed path in a plane perpendicular to I. As shown

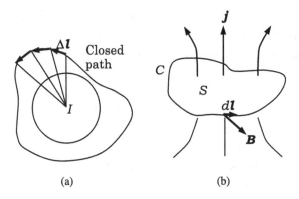

(a)   (b)

Fig. 4.21.

in Fig. 4.21(a), we divide the closed path into several small segments of length Δl. In general, it is not necessary for the direction of $\boldsymbol{B}$, the magnetic field caused by I, to coincide with that of Δl, a small-segment vector of the closed path (Δl is shown in Fig. 4.21(a)). Now, we replace the left-hand side of Eq. (4.21) by the sum of inner products of $\boldsymbol{B}$ and Δl. Further, we extend the current of the right-hand side of Eq. (4.21) to the total current I through the area enclosed by the path. The law obtained in such a way is known as **Ampere's law** (see Fig. 4.21(b)).

Ampere's law is expressed, for any closed path C, as

$$\sum_C \boldsymbol{B} \cdot \Delta l = \oint_C \boldsymbol{B} \cdot dl = \mu_0 I. \tag{4.22}$$

The integral in Eq. (4.22) is called the **line integral** of the magnetic field along the closed path C.

Example 4.7. A flat-plane conductor that is infinitely wide carries a uniform current per unit width, $\boldsymbol{j}$ ($|\boldsymbol{j}| = j$), which produces a uniform magnetic field, $\boldsymbol{B}$, parallel to the plane and perpendicular to the current. Consider a closed loop as shown in Fig. 4.22. The shape of the loop is rectangular and the rectangular plane surrounded by the loop is normal to the current. The lengths of the upper and lower edges are L. By using Ampere's law (Eq. (4.22)), find the magnitude of the magnetic field at the upper and lower edges.

Solution

The line integration of the magnetic field along the loop gives $2BL$, and the current passing through the rectangular plane is equal to jL.

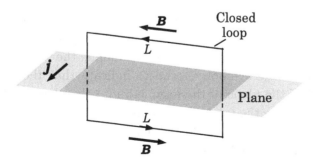

Fig. 4.22.

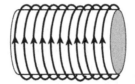

Fig. 4.23.

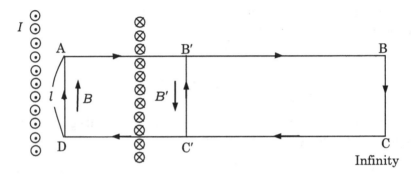

Fig. 4.24.

So, from Eq. (4.22), we have

$$2BL = \mu_0 jL \quad \therefore B = \frac{\mu_0 j}{2}. \qquad \blacksquare$$

Example 4.8. As shown in Fig. 4.23, the magnetic field caused by a current, I, in an infinitely long solenoid wound on a hollow cylinder is

$$B = \mu_0 nI \quad \text{inside the coil}, \qquad (4.23)$$

$$B = 0 \qquad \text{outside the coil}, \qquad (4.24)$$

where n is the number of turns per unit length of the coil.

Derive Eqs. (4.23) and (4.24) by using Ampere's law and Fig. 4.24, which shows a cross section containing the central axis of the coil. Here, the symbols $\odot$ and $\otimes$ indicate that the current is flowing out of the plane of the page and into the plane of the page, respectively. Since the solenoid is infinitely long, we can assume that both the magnetic fields inside and outside of the coil are parallel to the center axis of the solenoid and that the magnitude of the magnetic field at a point is dependent only on the distance from the conducting wire of the coil to the point. Moreover, take the magnitude of the magnetic field at an infinitely large distance away from the coil as zero.

Solution

As shown in Fig. 4.24, we apply Ampere's law to a closed rectangular loop A→B→C→D→A. Here, edges AB and CD are sufficiently long, and $BC = DA = l$. As the edges AB and CD are perpendicular to the central axis of the coil, the magnetic field is perpendicular to AB and CD. Therefore, the line integral from A to B and that from C to D are zero:

$$\int_{A\to B} B dl = \int_{C\to D} B dl = 0.$$

Further, as the edge BC is infinitely far from the coil, we may take $B = 0$ at any point on BC. The edge DA is parallel to the coil axis. So, the magnitude of the magnetic field on this edge is constant, and we have

$$\oint B dl = \int_{D\to A} B\, dl = Bl.$$

Now, as the number of turns of a solenoid of length l is nl, the current passing through the rectangular loop A→B→C→D is $nl \cdot I$. From Ampere's law (Eq. (4.22)), we obtain

$$Bl = \mu_0 nl I \quad \Rightarrow \quad B = \mu_0 n I.$$

From this result, we find that only if the edge DA is parallel to the central axis of the coil and is within the coil, the magnitude of the magnetic field is independent of the position of DA and is given by Eq. (4.23).

Next, we find the magnetic field outside the coil. To do this, we apply Ampere's law to the rectangular closed loop B′→B→C→ C′→B′. Because the edge C′B′ is parallel to the axis of the coil, the magnitude of the magnetic field on this edge, B', is constant. Thus, because the current through the closed loop B′→B→C→C′→B′ is zero, we can write

$$\oint B dl = \int_{C' \to B'} B dl = -B'l = 0 \quad \Rightarrow \quad B' = 0.$$

Also, since the distance from the wire wound around the coil to C'B' is arbitrary, the magnetic field outside the coil should be zero.

■

4.4.3. *The Lorentz force*

In a uniform magnetic field of magnitude B, the magnitude of the magnetic force, F, that acts on a point charge of q moving perpendicularly to the magnetic field at a speed of v is

$$F = qvB.$$

Suppose the point charge is positive, $q > 0$. Then, as shown in Fig. 4.25, if we let the left thumb, the index finger, and the middle finger open at right angles to one another, and let the direction of middle finger coincide with that of the velocity of the charge and the direction of index finger with that of the magnetic field, then the direction of the force is that of thumb.

In general, if a point charge of q moves at a velocity of $\boldsymbol{v}$ in a magnetic field of $\boldsymbol{B}$, then the following magnetic force, $\boldsymbol{f}$, acts on the

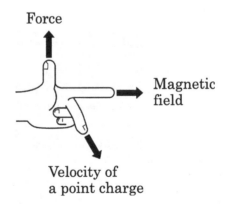

Force

Magnetic field

Velocity of
a point charge

Fig. 4.25.

charge:

$$f = qv \times B. \tag{4.25}$$

Here, $v \times B$ represent a vector product of v and B, (vector product was introduced in Sec. 2.6). The force given by Eq. (4.25) is called the **Lorentz force**.

When a charged particle with a charge of q moves at a velocity of v in the space where both an electric field of E and a magnetic field of B are present, both fields exert an electromagnetic force of $f = q(E + v \times B)$ on the particle. This force is also called the Lorentz force, in a broad sense.

The strength of a current is equal to the quantity of electricity passing through a section of a conducting wire per unit time. When the charge per unit length, λ, is uniform along a straight wire and the charges in the wire move at a velocity of v, the strength of the resulting current, I, is

$$I = \lambda v. \tag{4.26}$$

Consider a conductor carrying a current in a magnetic field. A magnetic force acts on every charge moving in the current-carrying conductor. As a consequence, we can regard this force as acting on the conductor.

Example 4.9. As shown in Fig. 4.26, a straight conductor of length a carries a current of I in a magnetic field of B perpendicular to the conductor. Show that the magnitude of the force exerted on the

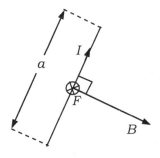

Fig. 4.26.

conductor is

$$F = BaI. \tag{4.27}$$

Solution

From Eq. (4.26), the charge contained in the conductor of length a is $\lambda a = aI/v$, and then, from Eq. (4.25), the magnitude of the magnetic force exerted on the conductor of length a by the magnetic field is

$$F = \frac{aI}{v} \times v \times B = BaI. \tag{4.28}$$

∎

Example 4.10. As shown in Fig. 4.27, a positive point charge, q, is moving at a velocity, v, in the same direction as a straight current, I. The distance between the current I and the charge q is r. Find the magnitude and the direction of the force exerted on q by the magnetic field induced by the current I.

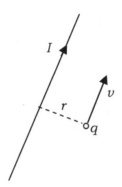

Fig. 4.27.

Solution

As the magnitude of the magnetic field at a distance r from a current I is

$$B = \frac{\mu_0 I}{2\pi r}, \tag{4.29}$$

the magnitude of the Lorentz force exerted on the charge is

$$f = qvB = \mu_0 \frac{qvI}{2\pi r}. \tag{4.30}$$

The force is perpendicular to and points toward the current. ∎

From the previous example, we see that a force acts between two parallel currents. As shown in Fig. 4.28, a current of $I_2 = I$ is flowing in (a straight) conductor 2. In (another straight) conductor 1, separated by a distance r from conductor 2, a uniform positive charge per unit length, λ, is distributed and is moving at a speed, v, in the same direction as I_2. Then the current in conductor 1, $I_1 = \lambda v$, flows in the same direction as I_2.

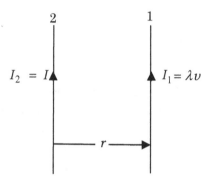

Fig. 4.28.

Now, by replacing q with λ in Eq. (4.30), we obtain the magnitude of the force exerted on I_1 per unit length as

$$f = \mu_0 \frac{I_1 I_2}{2\pi r}. \tag{4.31}$$

The force between I_1 and I_2 is attractive when they are in the same direction, and repulsive when they are in the opposite directions.

Equation (4.31) is utilized for the definition of the unit of current as previously presented in Sec. 4.1. That is, when we set $I_1 = I_2 = I$ and $r = 1\,\text{m}$ to have $f = 2 \times 10^{-7}\,\text{N}$, we define the current I as $1\,\text{A}$. By this definition, we are led to choose the value for μ_0 to be $4\pi \times 10^{-7}\,\text{N/A}^2$ from Eq. (4.31).

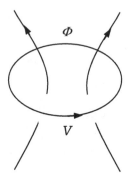

Fig. 4.29.

4.4.4. *Electromagnetic Induction and Self-Inductance*

The law of electromagnetic induction

As presented in Sec. 4.2, the magnetic flux (expressed by the number of magnetic field lines) through a coil changes when we move a magnet either toward or away from the coil. Then, an electromotive force (emf) is induced and an **induced current** flows in the coil. This phenomenon is called **electromagnetic induction**. It also occurs when we move a coil either toward or away from a fixed magnet. This phenomenon is generally stated as follows.

As shown in Fig. 4.29, let Φ be the magnetic flux through a coil of a single closed loop and V be the induced emf. Let us define the induced emf to be positive when its direction matches the winding direction of a right-handed screw pointing in the positive direction of the magnetic flux. Then, we have

$$V = -\frac{d\Phi}{dt}. \qquad (4.32)$$

This is called the **law of electromagnetic induction** or **Faraday's Law**.

Suppose we move the magnet while keeping the coil stationary. Then the free electrons in the conducting wire of the coil are stationary, and the magnetic field does not exert the Lorentz force on them. Nevertheless, a force acts on the electrons in the wire so as to make them move along the wire. Hence, we are forced to consider that there is an electric field in the wire. This field is called the **induced**

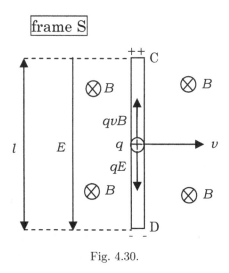

Fig. 4.30.

electric field. The induced electric field can be generated without a coil whenever the magnetic flux through a closed loop changes.

The Lorentz force and induced electromotive force

There is a phenomenon similar to the law of electromagnetic induction: it is the emf induced in a moving conducting wire in a magnetic field. This emf can be explained by the Lorentz force acting on the moving charge.

As shown in Fig. 4.30, in a reference frame, S, a conductor rod, CD, moves at a velocity, v, in a magnetic field, B. Here, the rod, the velocity and the magnetic field are perpendicular to one another. We assume there are freely moving positive charges in the rod. When a positive charge, q, is moving at the same velocity v as the rod (perpendicularly to the magnetic field), the magnetic field exerts the Lorentz force of magnitude qvB on q in the direction of D→C along the rod. Then, positive and negative charges continue to accumulate at end C and end D, respectively. As a result, an electrostatic field, E, is created in the direction of C→D within the rod. After a certain time, the Lorentz force on the charges balances with the force exerted by the electrostatic field and the charges finally become stationary. Using the balance of forces, we have, E, the magnitude of the electric field as

$$qvB = qE \quad \Rightarrow \quad E = vB \tag{4.33}$$

Let the length of the conductor rod be l. The potential at end C is higher than that at end D, and the potential difference, V, is

$$V = El = vBl. \tag{4.34}$$

In this way, an **induced emf** (which gives rise to the potential difference in Eq. (4.34)) **is caused by the Lorentz force in the conductor rod**. As the area swept by the conductor rod in a unit time is vl, **the magnitude of the induced emf is equal to the magnetic flux that the conductor rod crosses in a unit time**, i.e., $V = vBl$.

In comparison, in the reference frame that moves with the conductor rod, S′, the positive charges in the rod remain at rest, and therefore, the magnetic field does not exert the Lorentz force on the charges, as shown in Fig. 4.31. Nevertheless, the force directed from D to C acts on the positive charge q. So, there must be an electric field induced in the direction D→C in the reference frame S′. Since the magnitude of this force is equal to that of the Lorentz force

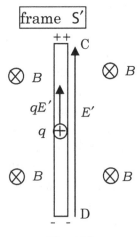

Fig. 4.31.

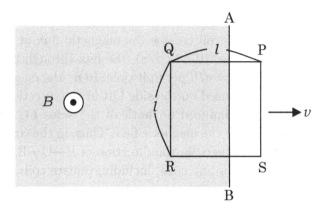

Fig. 4.32.

on q, the magnitude of the induced electric field, E', is given by

$$qE' = qvB \quad \Rightarrow \quad E' = vB.$$

The magnitude of the induced electric field, E', is equal to the electrostatic field, E, given by Eq. (4.33), but is opposite in direction.

As shown in Fig. 4.32, consider a motion of a square coil, PQRS, with sides each of length l. Suppose a uniform magnetic field directed out of the plane of the page is applied to the region on the left side of the line AB. The magnitude of the field is B. The magnetic field is not applied to the region on the right side of the line AB. Now, the coil starts moving rightward at a constant speed, v, then the Lorentz force of magnitude $f = qvB$ acts on a positive charge of q on the side QR in the direction Q→R. Hence, an emf is induced in the direction P→Q→R→S→P in the coil.

Example 4.11. Suppose at time $t = 0$, we put the side PS of the coil in Fig. 4.32 on the line AB. Let the magnetic flux through the coil at time $t(0 < t < l/v)$ be $\Phi(t)$. Show that the emf induced in the direction P→Q→R→S→P in the coil is

$$V = -\frac{d\Phi}{dt}. \tag{4.35}$$

Solution

Since the side QR of the coil crosses the magnetic flux at a rate of vBl per unit time at time $t(0 < t < l/v)$, the flux through the coil is reduced at a rate of $-\frac{d\Phi}{dt} = vBl$ per unit time. Here, the magnitude of the emf, $V = vBl$, is induced on the side QR in the direction Q→R. In contrast, emf is not induced on both of the sides PQ and RS, because they do not cross the magnetic flux. Thus, in the coil, an emf of $V = -\frac{d\Phi}{dt} = vBl$ is induced in the direction of P→Q→R→S→P.

    This relation holds for any coils, including square coils.   ■

Self-induction

When a current, I, in a coil varies with time, an emf is induced so as to counteract the change in the current. This emf can be expressed as

$$V = -L\frac{dI}{dt}. \tag{4.36}$$

    The emf given by Eq. (4.36) is called the **self-induced emf**. In Eq. (4.36), the coefficient L represents the magnitude of self-induction, and is called the **self-inductance**.

    As shown in Fig. 4.33, when an external flux, Φ, through a coil of a self-inductance L and of a resistance R varies with time, from Eq.(4.32) and Eq. (4.36), Kirchhoff's loop rule applied to the coil

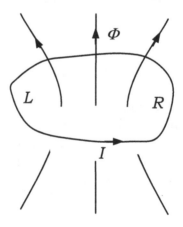

Fig. 4.33.

circuit is

$$-\frac{d\Phi}{dt} - L\frac{dI}{dt} = RI. \tag{4.37}$$

A time-dependent current is induced in the circuit, following Eq. (4.37). This equation is called the **circuit equation**. Here, the current has a sign. When the current I is positive, the coil carries the current in the rotational direction of a right-handed screw advancing toward the direction of the magnetic flux Φ, and when it is negative, the coil carries the current in the opposite direction.

Advanced Problems

Problem 4.4. The law of Bio and Savart

The battery that Volta invented in 1800 contributed a great deal to the advances of modern science. It gave scientists the gift of electric current. Twenty years later, Oersted of the University of Copenhagen, discovered that a current exerts force on a compass needle, and he reported this in a small article in June 1820. His discovery initiated a progress toward the unified concept of electricity and magnetism. In September of the same year, Arago reported Oersted's findings at a conference of the Institute of France. One week later, Ampere announced that he had found that two currents flowing in the same direction in two straight parallel wires attract each other, and that those flowing in the opposite directions repel each other. Seven weeks later, at the end of October of the same year, Biot and Savart announced their discovery of a law of the force that a linear electric current exerts on magnetic poles.

Figure 4.34 shows an overview of Biot and Savart's experiment. Electric current I is flowing along a long straight conducting wire in a direction, perpendicular to the paper. AB is a compass needle placed parallel to the paper. The needle is supported at its center P, around which it can rotate. The mass of the compass needle is m and its length is $2l$. Hereafter, we refer to "this needle" as "the compass needle". A magnet is placed close by in order to cancel the geomagnetic influence. Biot and Savart presumed the following.

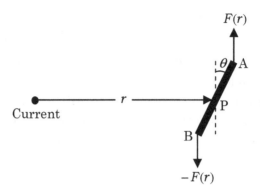

Fig. 4.34.

"The force exerted on a magnetic pole at point P is perpendicular to the conducting wire, and points in a direction tangential to a circle whose center is at the conducting wire. The strength of this force is proportional to the magnitude of current I and depends on the distance r between the linear current and the magnetic pole."

Let us express the force by $F(r)$ to indicate explicitly that it depends on r. When the compass needle tilts a little, as shown in Fig. 4.34, the distances between the current and the two ends, A, B, of the compass needle are $r_A = \sqrt{r^2 + l^2 + 2rl\sin\theta}$ and $r_B = \sqrt{r^2 + l^2 - 2rl\sin\theta}$, respectively. If l is substantially smaller than r, these distances can be approximated by $r_A = r$ and $r_B = r$. We set the compass needle such that the force exerted on point A is $F(r)$, and that on point B is $-F(r)$, both forces being in a direction tangential to the circle with its center at the conducting wire (Fig. 4.34).

Before applying an electric current, we tilt the compass needle by a small angle, θ, from the direction tangential to a circle whose center is at the conducting wire. When a current I flows through the wire, the compass needle vibrates around point P like a pendulum. Let us find how its vibration period t depends on m, l and $F(r)$.

(1) Let the period t be written as $t = hm^a l^b [F(r)]^c$. Determine a, b and c by a dimensional analysis to complete this equation for t, where h is a dimensionless constant.

(A physical quantity has fundamental dimensions, such as mass [M], length [L], and time [T], or combinations of these fundamental dimensions, such as velocity $[LT^{-1}]$, acceleration $[LT^{-2}]$, etc. Consideration of dimensions helps us determine the relation between physical quantities. For example, let us take a pendulum with a small mass m suspended by a string of length l. This system involves only three physical quantities: m, l and the gravitational acceleration g, whose dimensions are $[L]$, $[M]$, and $[LT^{-2}]$, respectively. Since the dimension of the period is $[T]$, the period can be assumed to be proportional to $\sqrt{\frac{l}{g}}$. This gives the period $= s\sqrt{\frac{l}{g}}$, where s is a constant. This type of analysis is called the dimensional analysis.)

(2) Let t_1 be the vibration period of the compass needle when the magnitude of the electric current is I, and let t_2 be the period when the current magnitude is $2I$. Assuming that the values other than the current are the same in both cases, find the time ratio $\frac{t_2}{t_1}$.

Biot and Savart varied the distance r between the linear current and the compass needle to measure the time interval of ten vibration periods for each r value. Table 4.1 shows their data.

(3) Assume $F(r) = ar^n$, where a is a constant. Which value of n, out of $n = 1, 0, -1$ and -2, makes this equation fit best to the data in Table 4.1?

This experiment showed how the force exerted on a compass needle at distance r from a linear electric current I depends on r, and that the force works in a direction tangential to a circle whose center

Table 4.1. Biot and Savart's experimental data.

r (mm)	Time interval of ten periods observed (sec)
15	30.00
20	33.50
40	48.85
50	54.75
60	56.75
120	89.00

is at the linear current. However, this experiment was not sufficiently accurate. Biot presumed, from calculus theory, that the conducting wire consisted of tiny elements and that the sum of the forces that each element exerted on the compass needle represented the force exerted by the whole wire. Based on a physical and mathematical consideration, he obtained the following equation:

$$f(r) = k\frac{I}{r}(-\cos\alpha + \cos\beta), \tag{4.38}$$

where $f(r)$ is the magnitude of the force that the wire element CD, in Fig. 4.35, exerts on the magnetic pole at point P (k is a constant).

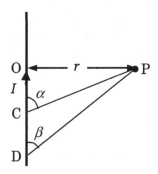

Fig. 4.35.

(4) Using Eq. (4.38), express $F(r)$ in terms of k, I and r. Here $F(r)$ is the magnitude of the force exerted on the magnetic pole in Fig. 4.34.

(5) Let $F_M(r)$ be the force that the half straight line below point M in Fig. 4.36 exerts on a magnetic pole at point P. Express $F_M(r)$ in terms of k, I, r and θ, using Eq. (4.38).

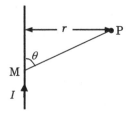

Fig. 4.36.

In order to confirm the correctness of Eq. (4.38), Biot came up with the ingenious idea of using a V-shaped current (see Fig. 4.37) instead of a linear one. He noticed that the force a V-shaped current exerts on a magnetic pole is equivalent to the force that a linear current, a part of which is missing, exerts on the same magnetic pole. Although a linear current, a part of which is missing, does not exist practically, its theoretical equivalent is given by a V-shaped current.

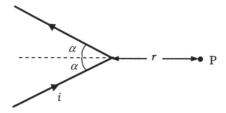

Fig. 4.37.

(6) Let $F'(r)$ be the force that the V-shaped current in Fig. 4.37 exerts on a magnetic pole at point P. Express $F'(r)$ in terms of α, i, r and k.

(7) Let t be the vibration period when the compass needle vibrates around point P in Fig. 4.34, and let t' be the vibration period when the needle vibrates around point P in Fig. 4.37. Express the ratio $\frac{t}{t'}$ in terms of parameters which are necessary in α, i, r, I and k.

Biot measured the vibration periods for various magnitudes of current, various distances between the magnetic pole and the current, and various angles of the V-shaped conducting wire, and showed that his theory is correct by comparing the ratios of these periods with his theoretical values.

In the 1820s, charged particles had not yet been discovered, nor had it been discovered that a current is a flow of charged particles. Because scientists had not yet arrived at the concept of magnetic field at that time, to Biot and other scientists of that era, his result was simply a law of the force that a current exerts on a magnetic pole, and not a law of the magnetic field generated by an electric current.

It was at the end of the 19th century that scientists established the principle of action through medium, which states that a current is a flow of charged particles and that a current forms around itself a magnetic field which affects magnetic poles and currents. Eventually, the Biot-Savart law came to be understood as, in fact, a law of the magnetic field generated by an electric current.

(the 2nd Challenge)

Solution

(1) Given mass, length, and force, solve for time. Then, you need to know the dimension of force.

$[t] = [\text{T}] = [m^a][l^b][F^c] = [\text{M}^a][\text{L}^b][(\text{MLT}^{-2})^c]$ gives $a + c = 0$, $b + c = 0$, and $-2c = 1$. Solving these equations, we obtain

$$a = \frac{1}{2}, \quad b = \frac{1}{2}, \quad c = -\frac{1}{2}.$$

The period is given by

$$t = h\sqrt{\frac{ml}{F(r)}}.$$

(2) From the result of part (1), we know that $F(r)$ is in the denominator in the expression for the period. Since $F(r)$ is proportional to the current, we have

$$\frac{t_{2I}}{t_I} = \sqrt{\left(\frac{1}{2I}\right) \Big/ \left(\frac{1}{I}\right)} = \frac{1}{\sqrt{2}}.$$

(3) Use $F(r) = ar^n$ in the result of part (1). The exponent n cannot be positive, since the period increases as r increases.

  If we let $n = -1$, then $t = s\sqrt{r}$ (s is a constant). Substitution of the values, $r = 15$ and $t = 30$, into this equation gives $s = 7.746$. Then, for $s = 7.746$ and $r = 20, 40, 50, 60, 120$ we obtain $t = 34.64, 48.99, 54.77, 60.00, 84.85$, respectively. These values agree favorably well with the experimental data. If $n = -2$, then $t = sr$. Therefore, substitution of $r = 15$ and $t = 30$ gives $s = 2$. Then for $s = 2$ and $r = 20, 40, 50, 60, 120$ we get $t = 40, 80, 100, 120, 240$, respectively. These values do not fit the experimental data.

It is easy to see that t goes farther away from the experimental data for $n = -3$, $n = -4$, etc. Therefore, the answer is $n = -1$.

(4) Applying $\alpha \to \pi$ and $\beta \to 0$ to Eq. (4.38), we obtain

$$F(r) = \frac{2kI}{r}.$$

(5) Applying $\alpha \to \theta$ and $\beta \to 0$ to Eq. (4.38), we obtain

$$F_\mathrm{M}(r) = \frac{kI}{r}(1 - \cos\theta).$$

(6) Subtracting the force that would have been produced by the missing part of the linear current from the force exerted by the entire linear current, we obtain the required value. Note that the distance between the imaginary linear current and point P is $r \sin\alpha$, which gives

$$F'(r) = \frac{2ki}{r \sin\alpha} - \frac{ki(-\cos(\pi - \alpha)) + \cos\alpha}{r \sin\alpha}$$

$$= \frac{2ki}{r \sin\alpha}(1 - \cos\alpha) = \frac{2ki}{r}\tan\frac{\alpha}{2}.$$

(7) Using the results of parts (4) and (6), we obtain

$$\frac{t}{t'} = \sqrt{\frac{\frac{ml}{2kI/r}}{\frac{ml}{(2ki/r)\tan(\alpha/2)}}} = \sqrt{\frac{i}{I}}\tan\frac{\alpha}{2}. \qquad \blacksquare$$

Problem 4.5.   The propagation of electromagnetic waves

Visible light, radio wave, X-ray, etc. belong to electromagnetic waves, which consist of waves of electric and magnetic fields, and propagate in vacuum at a constant speed $c \approx 3.0 \times 10^8\,\mathrm{m/s}$. The existence of the electromagnetic waves is theoretically predicted by J. C. Maxwell in the middle of the 19th century. Here, we will discuss Maxwell's equations.

Imagine a Cartesian coordinate system and choose the x-axis in the direction where an electromagnetic wave propagates. Then, the electric field E and the magnetic field B are perpendicular to each other, and are parallel to the y-axis and the z-axis, respectively (see Fig. 4.38). In the plane parallel to $y-z$ plane, both E and B are uniform and depend on position x and time t.

I The law of electromagnetic induction in a small area

We take a small rectangular loop abcd of a small width Δx and a small height h in the $x-y$ plane that contains point $\mathrm{P}(x,,0,0)$, as shown in Fig. 4.39. Let the electric field at a point $(x,0,0)$ be E and that at a point $(x + \Delta x, 0, 0)$ be $E + \Delta E$. And further, suppose the magnitude of the magnetic field B through abcd is uniform and that the magnetic field B changes to $B + \Delta B$ during a small time interval Δt.

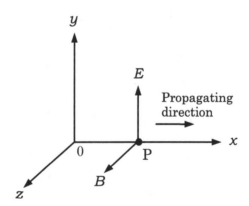

Fig. 4.38.

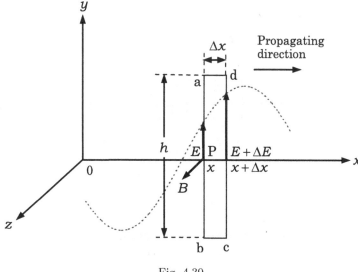

Fig. 4.39.

(1) Considering abcd a rectangular coil of a single turn, we apply the law of electromagnetic induction to this coil. Derive the following relation

$$\frac{\Delta E}{\Delta x} = -\frac{\Delta B}{\Delta t}. \qquad (4.39)$$

II Maxwell–Ampere's law

In the following sentence, enter appropriate equations in the boxes of $\boxed{(2)}$ and $\boxed{(3)}$.

Let a current i flow in a circuit with a capacitor of two parallel circular plates of area S as shown in Fig. 4.40. Any real current does not flow between the two plates of the capacitor, but Maxwell considered that a magnetic field B is generated around the capacitor as well as around the conductor wire of the circuit. Following this consideration, he supposed something equivalent to the current i flows between the plates. Assume that the area of the plates is sufficiently large and the space between the plates is very narrow. Then a uniform electric field perpendicular to the plates is produced between them.

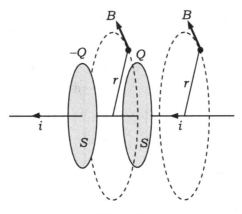

Fig. 4.40.

Now, let the vacuum permittivity be ε_0 and charges stored on the two plates be Q and $-Q$. Then an electric field E produced between the plates of the capacitor is given by $\boxed{(2)}$. Since the current i is the charge flowing into the capacitor per unit time, it is expressed in terms of the time derivative of the electric field, $\frac{dE}{dt}$, as

$$i = \boxed{(3)}. \tag{4.40}$$

Any real current does not flow between the plates, but the electric field between the plates varies with time. Maxwell considered that the electric field varying with time accompanies a magnetic field and called the physical quantity expressed by the right-hand side of Eq. (4.40) the **displacement current**.

Replacing the real current i by the displacement current i_D in Ampere's law, we have the following relation:

$$\oint B dl = \mu_0 i_D = \mu_0 \times \boxed{(3)}, \tag{4.41}$$

where μ_0 is the vacuum permeability. Equation (4.40) shows that when an electric field is varying with time, a magnetic field is induced. This is called **Maxwell–Ampere's law**.

III Maxwell–Ampere's law in small region

Let the magnetic field at the point x be B and that at the point $x+\Delta x$ be $B + \Delta B$. As shown in Fig. 4.41, we take a small rectangular loop

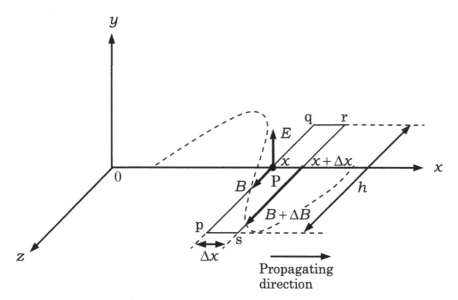

Fig. 4.41.

pqrs with a small width Δx and a small depth h in the $z-x$ plane that contains point P. Assume that the electric field through pqrs is spatially uniform, but depends on time and varies from E to $E+\Delta E$ during a small time interval Δt.

(4) By applying Maxwell-Ampere's law, derive the following relation:

$$\frac{\Delta B}{\Delta x} = -\varepsilon_0 \mu_0 \frac{\Delta E}{\Delta t}. \tag{4.42}$$

Since both E and B are functions of x and t, using partial derivatives with respect to x and t, we replace $\frac{\Delta E}{\Delta x}$ and $\frac{\Delta B}{\Delta t}$ in Eq. (4.39) and $\frac{\Delta E}{\Delta t}$ and $\frac{\Delta B}{\Delta x}$ in Eq. (4.42) by $\frac{\partial E}{\partial x}$, $\frac{\partial B}{\partial t}$, $\frac{\partial E}{\partial t}$ and $\frac{\partial B}{\partial x}$, respectively. Then we have

$$\frac{\partial E}{\partial x} = -\frac{\partial B}{\partial t}, \quad \frac{\partial B}{\partial x} = -\varepsilon_0 \mu_0 \frac{\partial E}{\partial t}. \tag{4.43}$$

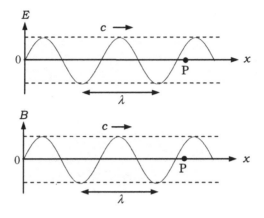

Fig. 4.42.

IV The propagation speed of an electromagnetic wave

We will derive the propagation speed c of an electromagnetic wave. As shown in Fig. 4.42, we assume that the electric field E and the magnetic field B are given as functions of x and t in the form of a progressive wave as follows:

$$E = E_0 \sin 2\pi \left(\frac{x}{\lambda} - \frac{t}{T} \right), \tag{4.44}$$

$$B = B_0 \sin 2\pi \left(\frac{x}{\lambda} - \frac{t}{T} \right). \tag{4.45}$$

Here, λ and T are, respectively, the wavelength and the period of E and B, and E_0 and B_0 are the amplitudes of E and B, respectively. We further note that both E and B are oscillating mutually in phase.

(5) Using Eq. (4.43), find the relation between the electric field E and the magnetic field B given by Eqs. (4.44) and (4.45), respectively. Further, show that the propagation speed of the electromagnetic wave is given by

$$c = \frac{1}{\sqrt{\varepsilon_0 \mu_0}}. \tag{4.46}$$

Substitution of the values of ε_0 and μ_0 into Eq. (4.46) yields $c \approx 3.0 \times 10^8 \, \text{m/s}$. We see that the visible light is a kind of electromagnetic wave, because the value of c coincides with the observed speed of light in vacuum.

(the 2nd Challenge)

Solution

(1) When the magnetic flux changes from Φ to $\Phi + \Delta\Phi$ during a small time interval Δt, the law of electromagnetic induction gives the emf V induced per a single turn as

$$V = -\frac{\Delta\Phi}{\Delta t}. \tag{4.47}$$

Since the area of the coil is $S = h\Delta x$, the magnetic flux through the coil is $\Phi = Bh\Delta x$, and so, we obtain

$$\frac{\Delta\Phi}{\Delta t} = h\Delta x\frac{\Delta B}{\Delta t}. \tag{4.48}$$

Let a unit charge go around the rectangular coil along c→d→a→b→c. The work done by the electric field is equal to the induced emf V. The works along the paths c→d, d→a, a→b and b→c are $W_{c \to d} = (E + \Delta E)h$, $W_{d \to a} = 0$, $W_{a \to b} = -Eh$ and $W_{b \to c} = 0$, respectively, and so, summing them up we have

$$V = W_{c \to d} + W_{d \to a} + W_{a \to b} + W_{b \to c}$$
$$= (E + \Delta E)h - Eh = h\Delta E. \tag{4.49}$$

From Eq. (4.47) through (4.49), we obtain Eq. (4.39).

(2) Since the charge per unit area of the plate of the capacitor is $\frac{Q}{S}$, the electric field E between the plates is

$$E = \frac{Q}{\varepsilon_0 S}. \tag{4.50}$$

(3) From Eq. (4.50), we can write a relation $\Delta E = \frac{\Delta Q}{\varepsilon_0 S}$. On the other hand, since the current is the charge flowing through the cross section of the conductor wire per unit time, we have

$$i = \frac{dQ}{dt} = \varepsilon_0 S\frac{dE}{dt}.$$

(4) Let us apply Maxwell-Ampere's law to the rectangular loop pqrs. The line integration of B along the loop q→p→s→r→q yields

$$\oint B\, dl = Bh - (B + \Delta B)h = -h\Delta B.$$

Since the area of the rectangle pqrs is $S = h\Delta x$, Eq. (4.41) then becomes

$$-h\Delta B = \mu_0\varepsilon_0 h\Delta x \frac{\Delta E}{\Delta t}.$$

(5) The propagation speed c of the electromagnetic wave is the velocity of the point where the phase of the wave is constant. Hence, letting $\frac{x}{\lambda} - \frac{t}{T}$ be a constant c_0 in Eqs. (4.44) and (4.45) and differentiating both sides of the equation, $\frac{x}{\lambda} - \frac{t}{T} = c_0$, with respect to t, we have the propagation speed c of the electromagnetic wave as

$$\frac{1}{\lambda}\frac{dx}{dt} - \frac{1}{T} = 0 \quad \Rightarrow c = \frac{dx}{dt} = \frac{\lambda}{T}. \tag{4.51}$$

Equation (4.51) is known as the fundamental equation of the wave.

From Eqs. (4.44) and (4.45), we get

$$\frac{\partial E}{\partial x} = \frac{2\pi}{\lambda}E_0 \cos 2\pi \left(\frac{x}{\lambda} - \frac{t}{T}\right),$$

$$\frac{\partial B}{\partial t} = -\frac{2\pi}{T}B_0 \cos 2\pi \left(\frac{x}{\lambda} - \frac{t}{T}\right).$$

Substituting these results to the first equation of Eq. (4.43), and using Eq. (4.51), we have

$$E_0 = \frac{\lambda}{T}B_0 = cB_0.$$

Substitution of this equation into Eq. (4.44) gives the relation between the electric field and magnetic field as

$$E = cB. \tag{4.52}$$

Further, substituting Eqs. (4.44) and (4.45) into the second equation of Eq. (4.43), and using Eq. (4.51), we have, in the same way as the above,

$$B = \frac{\lambda}{T}\varepsilon_0\mu_0 E = c\varepsilon_0\mu_0 E. \qquad (4.53)$$

Elimination of B from Eqs. (4.52) and (4.53) yields

$$E = c^2\varepsilon_0\mu_0 E,$$

from which Eq. (4.46) is derived. ∎

Problem 4.6. The motion of charged particles in a magnetic field

Dilute gases spreading in the region of the magnetosphere located above the ionosphere of the earth and in the interplanetary region of the solar system are, in many cases, in the state of plasmas (ionized gases) which consist of electrons and protons. These plasmas are moving under strong effects of magnetic and electric fields. Here, by considering motions of charged particles in the presence of magnetic and electric fields, we study the basis of the motion of the plasma. I We consider the motion of a particle of charge q and mass m in a uniform electric field $\boldsymbol{E} = (E, 0, 0)$ directed along the x-axis and a uniform magnetic field $\boldsymbol{B} = (0, 0, B)$ directed along the z-axis. Let the position of the particle be (x, y, z) and its velocity be (V_x, V_y, V_z), then we have the following set of equations:

$$m\frac{dV_x}{dt} = q(E + BV_y),$$
$$m\frac{dV_y}{dt} = -qBV_x, \qquad (4.54)$$
$$m\frac{dV_z}{dt} = 0.$$

Such equations are called the equations of motion; each left-hand side represents the acceleration times the mass of the particle, and each right-hand side represents the force acting on the particle from the electric field E and the magnetic field B. Here we assume that both E and B are independent of time. From the third equation of Eq. (4.54), we can easily see that in the z-direction the particle moves at a constant speed, so, in Sec. I, we will confine our considerations to the particle motion in the plane perpendicular to the z-axis.

At first, we set $E = 0$, then, from Eq. (4.54), we have

$$m\frac{dV_x}{dt} = qBV_y, \quad m\frac{dV_y}{dt} = -qBV_x. \tag{4.55}$$

(1) From Eq. (4.55), derive the following relations:

$$\frac{d^2V_x}{dt^2} = -\omega_c^2 V_x, \quad \frac{d^2V_y}{dt^2} = -\omega_c^2 V_y, \tag{4.56}$$

where

$$\omega_c = \frac{qB}{m}. \tag{4.57}$$

(2) Equation (4.56) shows that the charged particle performs a simple harmonic oscillation at angular frequency $|\omega_c|$ in both of the x and y directions. $|\omega_c|$ is called the **cyclotron angular frequency**. Letting $V_x = 0$ and $V_y = V_\perp > 0$ at $t = 0$, show that V_x and V_y at time t are, respectively, given by

$$V_x(t) = V_\perp \sin \omega_c t, \quad V_y(t) = V_\perp \cos \omega_c t$$

(3) Derive the position $(x(t), y(t))$ of the charged particle which was located at $(-\frac{V_\perp}{\omega_c}, 0)$ at the initial time $t = 0$. Further, show that the particle moves in a circle of radius given by

$$r_c = \frac{V_\perp}{|\omega_c|}. \tag{4.58}$$

This circular motion is called the **cyclotron motion**, and r_c is called the **cyclotron radius**. Note that by using the position coordinate (x, y), the velocity vector is given by

$$V_x(t) = \frac{dx}{dt}, \quad V_y(t) = \frac{dy}{dt}.$$

(4) The magnitude of the geomagnetic field depends on a position of the surface of the earth. At Tokyo, it is about 3×10^{-5} T. Calculate the cyclotron angular frequency of the electron at Tokyo.

(5) Calculate the cyclotron radius of the electron when its velocity is 3.2×10^7 m/s (which corresponds to the velocity of an electron accelerated by the electric voltage of about 3,000 V). Further, calculate the ratio of this cyclotron radius to the radius of the earth, 6.4×10^6 m.

The geomagnetic field, when observed in the earth scale length, depends on the position, but, if we observe it in the scale length of the cyclotron radius, which is much smaller than the radius of the earth, we can regard the charged particle motion as a circular motion in a uniform magnetic field.

Now, we consider the case that E is not zero in Eq. (4.54).

(6) Denoting $U_x = V_x$ and $U_y = V_y + \frac{E}{B}$, show that the equation of motion (4.54) can be rewritten as

$$m\frac{dU_x}{dt} = qBU_y, \quad m\frac{dU_y}{dt} = -qBU_x.$$

(7) By setting the initial condition as $U_x(0) = 0$, $U_y(0) = V_\perp > 0$, derive the expression for $V_x(t)$ and $V_y(t)$ in terms of $V_\perp$, ω_c, E, B and t.

(8) Using the above results, derive the particle position $(x(t), y(t))$ at time t under the initial conditions, $x = -\frac{V_\perp}{\omega_c}$ and $y = 0$ at $t = 0$. Draw the outline of the trajectory of its motion in both cases of $q > 0$ and $q < 0$, respectively.

(9) In part (8) the electric field was in the x-direction. Explain how the orbit looks like when the electric field is in the y-direction. Draw the outline of the trajectory of its motion in both cases of $q > 0$, and $q < 0$, respectively.

The above results show that in the presence of uniform magnetic and electric fields, which are perpendicular to each other, the motion of a charged particle in the plane perpendicular to the magnetic field is a superposition of the cyclotron motion and a uniform velocity motion in the direction perpendicular to both of the magnetic and electric fields. This motion is called the **$E \times B$ drift motion**.

(10) The plasma is a dilute ionized gas consisting of equal numbers of electrons and positive ions, being electrically neutral as a

whole. Suppose, to a plasma in a magnetic field, we apply an electric field perpendicular to the magnetic field, then explain how the motion of the plasma looks like as a whole.

II We next consider the motion of a charged particle helically winding around slightly curved magnetic field lines, as shown in Fig. 4.43. We choose the z-axis in the direction of the movement of the center of the cyclotron motion of the charged particle. The distribution of the magnetic field lines is symmetric around the z-axis. When the magnetic field consists of curved magnetic field lines as shown in Fig. 4.43, the magnetic field is not uniform along the z-axis. Then, the charged particle moves with acceleration in the z-direction.

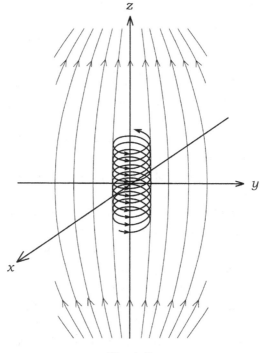

Fig. 4.43.

As the charged particle moves in the z-direction, the magnetic field B at the particle position varies with time, but if its variation during one period of the cyclotron motion is much smaller than the strength of B itself at that position, then, the motion in this field can approximately be described by a superposition of the cyclotron motion in the $x-y$ plane and the motion of the rotating center in the z-direction. Here, the angular frequency of the cyclotron motion varies with z, being written in terms of the local magnetic field, $B_z(Z)$, at the center of the circular motion as

$$\omega_c(z) = \frac{qB_z(z)}{m}. \tag{4.59}$$

In the treatment of Sec. I, V_z was constant as there was no force acted in the z-direction. Here, however, it depends on z, so that we have to consider an accelerated motion in the z-direction. For this purpose, we use the radius of the cyclotron motion, r_c, the relation (4.58) between the velocity of the circular motion $V_\perp > 0$ and the cyclotron frequency ω_c given by Eq. (4.59), and in addition the following two laws:

(a) Here, as the magnetic field variation is sufficiently slow, the force acting on the charged particle from the magnetic field can be regarded as being always directed to the center of the rotating motion. Then the law that the areal velocity is constant in rotating motion holds:

$$r_c V_\perp = C_1 \quad (C_1 : \text{ a constant}). \tag{4.60}$$

(b) As the force exerted by the magnetic field is perpendicular to the velocity of the charged particle, it does no work on the particle, so that the kinetic energy of the particle is constant, from which we have the following relation:

$$\frac{1}{2}m(V_z^2 + V_\perp^2) = C_2 \quad (C_2 : \text{ a constant}). \tag{4.61}$$

(11) Derive the expressions for C_1 and C_2 in Eqs. (4.60) and (4.61), respectively, in terms of $V_\perp(0)$, $V_z(0)$, $B_z(0)$, q and m.
(12) Show that, when B_z varies with the movement of the center of the rotating motion in the z-direction, $\frac{1}{2}mV_\perp(z)^2$ varies proportionally to $B_z(z)$.

(13) We introduce a function of z:

$$U(z) = \frac{mV_\perp(0)^2 B_z(z)}{2B_z(0)}.$$

Derive the following relation for the motion of the charged particle in the z-direction:

$$\frac{1}{2}mV_z(z)^2 + U(z) = C_2. \tag{4.62}$$

If we regard $U(z)$ in Eq. (4.62) as a potential energy at z, we can see that Eq. (4.62) is the law of the mechanical energy conservation of the charged particle for one dimensional motion in the z-direction. This means, when the magnetic field varies, a force acts on the charged particle in the direction of the magnetic field lines.

Hereafter, the magnetic field $B_z(z)$ is assumed to be an increasing function of $|z|$ having a minimum at $z = 0$, and is symmetric with respect to $z = 0$. The magnetic field shown in Fig. 4.43 is an example of such magnetic field configuration.

(14) Suppose there are two plates at $z = \pm L$, respectively, in the magnetic field configuration, as shown in Fig. 4.44. We assume that when the charged particle hits the plate, it is immediately absorbed and annihilated. We denote the ratio of the maximum value, $B_z(L)$, to the minimum value, $B_z(0)$, by M, i.e.,

$$M = \frac{B_z(L)}{B_z(0)}.$$

Derive the condition for $V_\perp(0)$ and $V_z(0)$ that the charged particle should move back and forth between two plates without hitting the plates, in terms of M.

Now, we consider the charged particle velocity at $z = 0$ in the three dimensional velocity space, (V_x, V_y, V_z). The result of the previous part (14) shows that in this three dimensional velocity space there is a cone-shaped region whose axis is the V_z-axis. This cone-shaped region in the velocity space is called the **loss cone**, inside

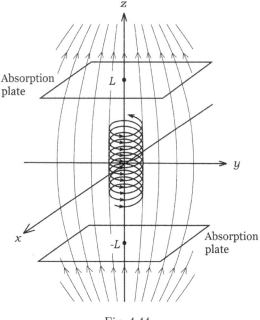

Fig. 4.44.

which all charged particles are lost by hitting the end plates. This is a characteristic property of the plasma enclosed in a magnetic configuration, such as the one in the geomagnetic region. Stated in another way, the charged particles outside the region of the loss cone in the velocity space are confined near the minimum point of the magnetic field. This is called the **magnetic mirror effect**.

(15) Draw the shape of the loss cone in the velocity space, and derive the value of $\sin\theta$ in terms of the ratio M, where apex angle of the cone is 2θ.

A group of charged particles (geomagnetic plasma) confined by the magnetic mirror effect of the geomagnetic field are, sometimes, exposed to and shaken by the plasma wind ejected from the sun. Then, some charged particles escape from the magnetic mirror, and they storm into the ionosphere near the South and North Poles, where magnetic field lines of the geomagnetic region descend. As a result, it is well known that auroras occur.

Notes: Electron mass: $m = 9.1 \times 10^{-31}$ kg

Elementary electric charge: $e = 1.6 \times 10^{-19}$ C

(the 2nd Challenge)

Solution

(1) By differentiating the first equation of Eq. (4.55) and using its second equation to the right hand side, we have

$$m\frac{d^2 V_x}{dt^2} = qB\frac{dV_y}{dt} = -\frac{q^2 B^2}{m}V_x.$$

Similarly, differentiation of the second equation of Eq. (4.55) yields

$$m\frac{d^2 V_y}{dt^2} = -qB\frac{dV_x}{dt} = -\frac{q^2 B^2}{m}V_y.$$

By using $\omega_c = \frac{qB}{m}$, we obtain Eq. (4.56).

(2) For the first equation of Eq. (4.56), $\frac{d^2 V_x}{dt^2} = -\omega_c^2 V_x$, we have the following general solution:

$$V_x(t) = a \sin \omega_c t + b \cos \omega_c t, \text{ where } a \text{ and } b \text{ are constants.}$$

From the initial condition that $V_x = 0$ at $t = 0$, we have $b = 0$. Then, from the equation of motion, we obtain

$$V_y = \frac{m}{qB}\frac{dV_x}{dt} = \frac{m}{qB}a\omega_c \cos \omega_c t = a \cos \omega_c t.$$

From $V_y = V_\perp$ at $t = 0$, we have $a = V_\perp$. Finally, we obtain $V_x = V_\perp \sin \omega_c t$, and $V_y = V_\perp \cos \omega_c t$.

(3) From $V_x = \frac{dx}{dt} = V_\perp \sin \omega_c t$, we have $x = x_0 - \frac{V_\perp}{\omega_c} \cos \omega_c t$, where x_0 is an integral constant. By the initial condition that $x = -\frac{V_\perp}{\omega_c}$ at $t = 0$, we have $x_0 = 0$, from which we obtain $x(t) = -\frac{V_\perp}{\omega_c} \cos \omega_c t$. Similarly, from $V_y = \frac{dy}{dt} = V_\perp \cos \omega_c t$, we have $y = y_0 + \frac{V_\perp}{\omega_c} \sin \omega_c t$, where y_0 is an integral constant. By the

initial condition that $y = 0$ at $t = 0$, we have $y_0 = 0$, from which we obtain $y(t) = \frac{V_\perp}{\omega_c} \sin \omega_c t$.

From these results, we have a relation, $x^2 + y^2 = (\frac{V_\perp}{\omega_c})^2$, which implies that the particle moves on a circle of radius $\frac{V_\perp}{|\omega_c|}$.

(4) From Eq. (4.57),

$$|\omega_c| = \frac{|q| B}{m} = \frac{1.6 \times 10^{-19}\,\mathrm{C} \times 3 \times 10^{-5}\,\mathrm{T}}{9.1 \times 10^{-31}\,\mathrm{kg}} = 5.3 \times 10^6\,\mathrm{rad/s}.$$

(5) By using the values of $V_\perp = 3.2 \times 10^7\,\mathrm{m/s}$, and $|\omega_c| = 5.3 \times 10^6\,\mathrm{rad/s}$, we get

$$r_c = \frac{V_\perp}{|\omega_c|} = \frac{3.2 \times 10^7\,\mathrm{m/s}}{5.3 \times 10^6\,\mathrm{rad/s}} = 6.0\,\mathrm{m}.$$

The ratio of the cyclotron radius to the radius of the earth, $R = 6.4 \times 10^6\,\mathrm{m}$, is

$$\frac{r_c}{R} = \frac{6.0\,\mathrm{m}}{6.4 \times 10^6\,\mathrm{m}} = 9.4 \times 10^{-7}.$$

(6) We substitute $V_x = U_x$ and $V_y = U_y - \frac{E}{B}$ into the first and the second equations of Eq. (4.54). Since $E + BV_y = BU_y$ and $\frac{dV_y}{dt} = \frac{dU_y}{dt}$, we have

$$m\frac{dU_x}{dt} = qBU_y, \quad m\frac{dU_y}{dt} = -qBU_x.$$

(7) By using part (2) and the initial conditions, we have

$$U_x(t) = V_\perp \sin \omega_c t, \quad U_y(t) = V_\perp \cos \omega_c t.$$

So, we obtain the following relations

$$V_x(t) = U_x(t) = V_\perp \sin \omega_c t,$$

$$V_y(t) = U_y(t) - \frac{E}{B} = V_\perp \cos \omega_c t - \frac{E}{B}.$$

(8) Integrating the results of the previous part (7) with respect to t, we have

$$x = x_0 - \frac{V_\perp}{\omega_c} \cos \omega_c t \quad (x_0 : \text{an integral constant})$$

$$y = y_0 + \frac{V_\perp}{\omega_c} \sin \omega_c t - \frac{E}{B} t \quad (y_0 : \text{an integral constant}).$$

Since $x = -\frac{V_\perp}{\omega_c}$ and $y = 0$ at $t = 0$, we have $x_0 = 0$, $y_0 = 0$. Then, we have

$$\underline{x = -\frac{V_\perp}{\omega_c} \cos \omega_c t}, \quad \underline{y = \frac{V_\perp}{\omega_c} \sin \omega_c t - \frac{E}{B} t}.$$

The outline of the trajectory of the cyclotron motion:
The sign of $\omega_c = \frac{qB}{m}$ coincides with that of q. When $\omega_c < 0$, it is easier that we show the outline, if we represent x and y in the following form:

$$x = \frac{V_\perp}{|\omega_c|} \cos |\omega_c| t, \quad y = \frac{V_\perp}{|\omega_c|} \sin |\omega_c| t - \frac{E}{B} t.$$

In a case of $V_\perp > \frac{E}{B}$, the direction of rotation of the cyclotron motion is clockwise when $q > 0$ ($\omega_c > 0$), and counterclockwise when $q < 0$, ($\omega_c < 0$), respectively. In either case the center of the rotation moves straight to the $-y$-direction at a uniform velocity. In a case of $V_\perp < \frac{E}{B}$, the y-component of velocity is always negative and the particle does not rotate. Outlines of the trajectories are shown in Fig. 4.45.

(9) When the electric field E is in the y-direction, from symmetry consideration outlines of the trajectories can be obtained by rotating the trajectories of part (8) by 90° counterclockwise around the z-axis. Namely, the direction of rotation of the cyclotron motion is the same as the case of E in the x-direction, and the motion of the center of rotation is a straight motion with uniform velocity in the x-direction for both cases of positive and negative charge. Outlines of the trajectories for the case $V_\perp > \frac{E}{B}$ are shown in Fig. 4.46 (the case of $V_\perp < \frac{E}{B}$ can, of course, be drawn in a similar way).

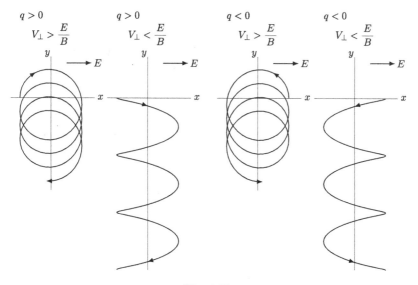

Fig. 4.45.

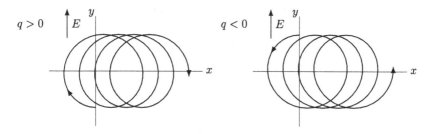

Fig. 4.46.

(10) By the results of part (7), we can see that the direction of the $E \times B$ drift, that is, the moving direction of the rotating center of the cyclotron motion, is the progressive direction of the right-handed screw from the E-direction to the B-direction. Further, the velocity of this drift is independent of the charge and the mass of the particle. Thus, we can say that a plasma consisting of electrons and positive ions moves as a whole to the direction of the $E \times B$ drift keeping its shape.

(11) From Eqs. (4.60) and (4.61), we have

$$C_1 = r_c V_\perp = \frac{V_\perp}{|\omega_c|} V_\perp = \frac{m}{|q|B_z} V_\perp^2 = \frac{m}{|q|B_z(0)} V_\perp(0)^2, \qquad (4.63)$$

$$C_2 = \frac{1}{2}m(V_z^2 + V_\perp^2) = \frac{1}{2}m(V_z(0)^2 + V_\perp(0)^2).$$

(12) From Eq. (4.63), we have

$$\frac{1}{2}mV_\perp(z)^2 = \frac{mV_\perp(0)^2}{2B_z(0)}B_z(z).$$

(13) The result of (12) can be written as $\frac{1}{2}mV_\perp(z)^2 = U(z)$, so, Eq. (4.61) becomes

$$\frac{1}{2}mV_z(z)^2 + U(z) = C_2.$$

(14) From parts (11) through (13), we have

$$\frac{1}{2}mV_z(z)^2 + \frac{1}{2}mV_\perp(0)^2\frac{B_z(z)}{B_z(0)} = \frac{1}{2}mV_z(0)^2 + \frac{1}{2}mV_\perp(0)^2.$$

We therefore obtain the relation

$$V_z(z)^2 = V_z(0)^2 + V_\perp(0)^2 - V_\perp(0)^2\frac{B_z(z)}{B_z(0)}.$$

In order that the charged particle turns its direction before hitting the plate, $V_z(z)$ should become zero in the region $|z| < L$. This requires the following relations:

$$V_z(0)^2 + V_\perp(0)^2 < V_\perp(0)^2\frac{B_z(L)}{B_z(0)} = V_\perp(0)^2 M,$$

from which we finally get the following condition:

$$\left|\frac{V_z(0)}{V_\perp(0)}\right| < \sqrt{M-1}.$$

(15) In the velocity space at $z = 0$, the region satisfying the inequality of the result of part (14), i.e., $|\frac{V_z(0)}{V_\perp(0)}| < \sqrt{M-1}$, is the space outside the cone shown in Fig. 4.47.

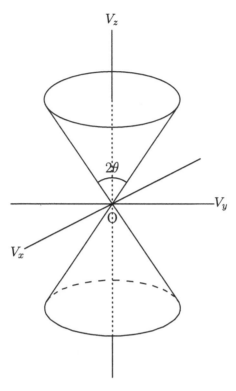

Fig. 4.47.

As the apex angle of the loss cone is 2θ, we have

$$\tan\theta = \frac{1}{\sqrt{M-1}}.$$

Namely,

$$M = 1 + \frac{1}{\tan^2\theta} = \frac{1}{\sin^2\theta},$$

$$\sin\theta = \frac{1}{\sqrt{M}}.$$

Chapter 5

Thermodynamics

Elementary Course

5.1. Heat and Temperature

When two systems are kept in thermal contact with each other for a sufficiently long time, they become a state of thermal balance and their states no longer change afterwards. Such a state is called a **thermal equilibrium state** where **the temperatures** of both systems are equal to each other. A physical quantity that assumes a definite value in a thermal equilibrium state is called a **quantity of state**. For example, the temperature, pressure, and volume of the gas are quantities of state.

5.1.1. *Empirical Temperature*

The temperature used in ordinary life is defined as follows.

When ice and water coexist in thermal equilibrium under atmospheric pressure, the temperature is set to be 0°C. When water and steam coexist, the temperature is set to be 100°C. Hence, by dividing the interval between the above two temperatures into one hundred equal parts, we define the temperature of each part as 1°C.

The ideal gas is used as a standard in dividing the interval into one hundred equal parts. The temperature defined in this way is called the **empirical temperature**.

5.1.2. *One Mole and Avogadro's Number*

One mole is defined by the amount of matter, which consists of the same number of particles as the number of atoms contained in 0.012 kilogram of carbon (mass number: 12). The number of atoms (molecules) contained in 1mol of the substance is called "**Avogadro's number**". It is given by the following number:

$$N_A = 6.02 \times 10^{23} \text{ 1/mol} \tag{5.1}$$

5.1.3. *Equation of State for Ideal Gas*

A gas is called an **ideal gas** when the constituent molecules are so far away from one another that interactions among molecules are negligible except for their collisions. The air under the atmospheric pressure can be regarded as an ideal gas because its density is small or the gas is dilute. In the ideal gas, the product of the pressure, p, and volume, V, is kept constant under a fixed temperature. This is called **Boyle's law**. It is established by experiments. The product pV is proportional to the molar number, n, of the gas molecules. Hence, **the absolute temperature**, T, is defined by a quantity that is proportional to pV of one mole ideal gas, and its coefficient is called the **gas constant**, being denoted by R. Thus, the following relation holds:

$$pV = nRT. \tag{5.2}$$

Equation (5.2) is the **equation of state for an ideal gas** we call the ideal gas equation. The unit of the absolute temperature is represented by K and its unit is defined by the relation, $1\,\text{K} = 1\,^\circ\text{C}$. Using an ideal gas, $0\,^\circ\text{C}$ is experimentally found to be $273.15\,\text{K}$. It is known that there is no essential difference between the absolute temperature defined empirically in this way and the **thermodynamic temperature** defined strictly by thermodynamics.

5.1.4. *Quantity of Heat and Heat Capacity*

Atoms and molecules are moving randomly even inside static matters. The higher the temperature is, the more rapid the motion of the particles is. This random motion of the particles is called the **thermal motion**. Change of the temperature of matter is due to a transfer of energy of thermal motion. This energy of thermal motion is called the **quantity of heat**. It is measured by the unit of work, J (Joule), because it is a kind of energy. The quantity of heat necessary for raising the temperature of water 1 g by 1°C is defined as 1 cal, and it is known that 1 cal = 4.2 J. **Heat capacity** is defined by the quantity of heat needed to raise the temperature of matter by 1 K. **Specific heat** is the quantity of heat needed to raise the temperature of matter of 1 kg by 1 K. Therefore, the quantity of heat, Q, which is needed for raising the temperature of matter with mass m and specific heat c by t is given by

$$Q = mct. \tag{5.3}$$

Elementary Problems

Problem 5.1. Properties of temperature

Read the following sentences (1)~(7) and check whether they are correct (Yes) or incorrect (No).

(1) When a ball made of copper (80°C) is soaked into the water (20°C) of the same mass as the ball contained in a heat insulating vessel, the water temperature becomes 50°C in thermal equilibrium.

(2) The wet bulb of a psychrometer indicates a temperature lower than that of the dry bulb, because the water evaporates from the surface of the wet bulb.

(3) Pumping air into a tire of a bicycle heats the cylinder of the bicycle pump. The reason is mainly not due to friction but due to adiabatic compression of the air in the cylinder.

(4) When an ideal gas of a given amount confined in a container is heated by keeping its volume constant, all the heat added is used to increase the internal energy.

(5) Heat quantity needed to raise the temperature of ideal gas of a given amount by 1°C is smaller under a constant pressure than under a constant volume.

(6) The absolute temperature of a gas is proportional to the average speed of the gas molecules.

(7) The pressure of the ideal gas in a container of constant volume is proportional to the average kinetic energy of the gas molecules.

(the 1st Challenge)

Answer (1) No (2) Yes (3) Yes (4) Yes (5) No (6) No (7) Yes

Solution

(1) When a matter absorbs heat, its temperature rises. Between the quantity of heat absorbed, Q, and the rise of temperature, ΔT, the relation $Q = C\Delta T$ holds (C is the heat capacity). It is known empirically that the larger the quantity of the matter is, the greater the heat capacity is. For example, the more the amount of water is, the longer the time needed for boiling is. The specific heat depends on the kind of matter. The system reaches thermal equilibrium at less than 50°C because the specific heat of copper is less than a half that of water.

(2) A psychrometer consists of two thermometers of the same type. The bulb of one of the thermometers is wrapped by wet cloth. Water evaporates from its surface when the humidity is low and as a result the temperature of the thermometer with wet cloth decreases. The humidity can be measured by comparing the temperatures indicated in the two thermometers. When, not only water, but any kind of liquid changes into gas, heat is deprived. It is called the heat of vaporization.

(3) When a gas is compressed without heat conduction, the temperature rises. The reason is that the work done for compressing the gas increases its internal kinetic energy. In contrast, the temperature falls with adiabatic expansion.

(4) Let Q be the absorbed heat and W be the work done on outer surroundings, then the increase of the internal energy ΔU is given by $\Delta U = Q - W$ according to the first law of thermodynamics. If the volume of the gas is kept constant, the system does no work on the outer surroundings. Therefore ΔU is equal to Q.

(5) The internal energy of an ideal gas is determined by the temperature alone. In any change of state of an ideal gas in a given amount, the increase of the internal kinetic energy

needed to raise the temperature by 1°C is identical. When the pressure is kept constant, the system does work on the outer surroundings due to the increase of the volume. By the first law of thermodynamics, $Q = \Delta U + W$, more heat is needed under constant pressure as compared with the case of constant volume to increase the same amount of the internal energy or the temperature.

(6) In gaseous states, molecules are disconnected from one another and move at high speeds. Their average kinetic energy is proportional to the absolute temperature. A molecule with mass m at velocity v has the kinetic energy, $\frac{1}{2}mv^2$, so the absolute temperature is not proportional to the velocity but to the square of the velocity.

(7) Let p be the pressure, T the absolute temperature, V the volume of an ideal gas of n moles, then by denoting the gas constant with R, we have the relation, $pV = nRT$. If the volume, V, is kept constant, the pressure, p, is proportional to the absolute temperature, T. From part (6), the pressure, p, is proportional to the average kinetic energy of the molecules.

(the 1st Challenge)

■

Problem 5.2. Potential energy and heat

A bag containing copper particles of 1kg is dropped at initial speed zero from a fixed point P to a rigid floor (the height of P is 1.5 m above the floor). This bag does not bound at all when it drops on and collides with the floor. By this collision, 70% of the kinetic energy of the particles within the bag is given to the particles as heat. How many times should you drop this bag in order to raise the temperature of particles by 5 K. Choose the most favorable answer.

The acceleration of gravity is 9.8 m/s². The specific heat of copper is 0.38 J/(g · K). Since the heat conduction of copper is well, you can assume that the heat given to the particles is uniformly delivered to all the particles without delay.

(a) 90 times   (b) 130 times   (c) 190 times   (d) 1800 times

(the 1st Challenge)

Answer (c)

Solution

A matter with mass m at height h has a potential energy, mgh. This potential energy changes into the kinetic energy by falling to the ground. The kinetic energy changes into molecule's vibration energy by collisions, which is the thermal energy of molecules.

Let n be the number of collisions. The energy given to molecules is $E = n \times 1 \times 9.8 \times 1.5 \times 0.70$ J. The quantity of heat required for raising the temperature by 5°C is $Q = 0.38 \times 1000 \times 5$ J. Equating E to Q, we obtain $n = 185 \approx \underline{190 \text{ times}}$ ■

Problem 5.3.   Change of the state of an ideal gas

The state of an ideal gas is changed as A → B → C → D → A. The relation between the pressure, p, and the volume, V, of the gas is shown in Fig. 5.1.

Among the five graphs shown in Fig. 5.2, which is the graph corresponding to the above change for the relation between the volume, V, and the temperature, T? Choose the most suitable one from the five options below.

(the 1st Challenge)

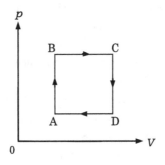

Fig. 5.1.

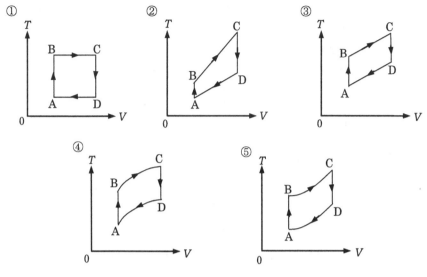

Fig. 5.2.

Answer ②

Solution

The ideal gas equation, $pV = nRT$, can be regarded as an equation for the relation among the quantities of p, V and T, where n and R are constant. If the values of two quantities out of three p, V and T, are given, the value of the remaining one is uniquely determined by the ideal gas equation. Keeping this in mind, we can consider that points on Fig. 5.1 ($p - V$ graph) represent the states of the gas and a transfer line on the graph represents the change of the state of the gas.

- A→B: The volume remains constant, hence the pressure, p, is proportional to the absolute temperature, T. The increase of p implies the increase of T under constant volume V.
- B→C: The pressure, p, remains constant, hence the volume, V, is proportional to the absolute temperature, T.

- C→D: The pressure, p, and the temperature, T, decrease under constant volume V.
- D→A: The volume, V, is proportional to the absolute temperature, T, whence V decreases as T decreases. The state D eventually returns to the starting point A.

  ② Line BC and line AD go through the origin and V is proportional to T.

  ③ Line BC and line AD do not go through the origin and V is not proportional to T.

(the 1st Challenge)

■

Problem 5.4.   Making water hotter than tea

As in Fig. 5.3, there are two heat insulating containers A and B and two other containers of high thermal conductivity C and D. Container A contains tea (80°C, the specific heat is equal to that of water) of $1\,L$. Container B contains water (20°C) of $1\,L$. The containers, C and D, can be put into the containers, A and B, such that the liquids inside A and B do not overflow. These containers can be used in the following ways.

(1) All the water in container B is poured into container C, and then container C is stored in A.

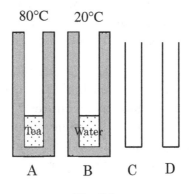

Fig. 5.3.

(2) Due to thermal conduction, the temperature of tea in container A later becomes equal to that of water in container C. Thus the following relation holds.

the quantity of heat emitted from container A

= the quantity of heat absorbed into container C

Here we ignored the quantity of heat used for the change of temperature of the container, as well as the heat emission to outer environment.

Can you make the temperature of water higher than that of tea, by using only thermal conduction processes with containers A $\sim$ D without mixing water and tea. If you can do it, explain the method. If you cannot do it, explain why.

Note that the following equation holds:

quantity of heat transferred

= amount of the matter $\times$ specific heat $\times$ change of temperature

(the 1st Challenge)

Answer We can do it. For example, there is a way below.

(a) Split the water in B half and half into C and D.
(b) C is stored in A, and then the temperature becomes 60°C. After that, the water is returned into B.
(c) D is stored in A and the temperature becomes 47°C. Then, the water is returned into B.
(d) The water in B becomes 53°C and the tea in A becomes 47°C.

We may also split the tea, instead of water. Moreover, it is not necessary to split the water strictly half and half.

Advanced Course

5.2. Kinetic Theory of Gases

Kinetic theory of gases is constructed with the aim to understand the nature of gases based on the microscopic motion of gas molecules. We treat the ideal gas as having the feature that the size of gas molecules can be neglected and there are no forces acting on them except for collisions among themselves. We treat the collisions, whether they are among molecules or between molecules and walls, as elastic collisions. Further no frictions are assumed to act between walls and molecules.

5.2.1. *Gas Pressure*

Consider n moles of an ideal gas at absolute temperature T contained in a cubic box with a length of each side, L, as shown in Fig. 5.4. The mass of a gas molecule is m and the number of gas molecules of n moles is N. The x-axis is vertical to the surface of the cube, S. The y- and z-axes are vertical to the x-axis. At a given moment, let the velocity of one of the gas molecules be $\boldsymbol{v} = (v_x, v_y, v_z)(v_x > 0)$. This molecule moves towards the surface S and hits this surface, then the x-component of its velocity changes from v_x to $-v_x$. So, the amount of impulse on the surface, S, by this collision is

$$|m(-v_x) - mv_x| = 2mv_x.$$

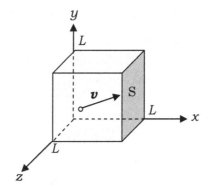

Fig. 5.4.

We assume that this molecule makes no collisions with other molecules, and it moves a distance v_x along the x-axis in unit time. Therefore, the number of collisions of this molecule with the surface S per unit time is $v_x/2L$. The impulse of this gas molecule on the surface per unit time can be regarded as an average force on the surface by this gas molecule. Hence, this force, $\langle f \rangle$, can be expressed as follows:

$$\langle f \rangle = \left\langle 2mv_x \frac{v_x}{2L} \right\rangle = \frac{m\langle v_x^2 \rangle}{L},$$

where $\langle f \rangle$ and $\langle v_x^2 \rangle$ denote the averages of f and v_x^2, respectively, over all the molecules.

Now, there are N molecules in this container, so the average force $\langle F \rangle$ on the surface by all the molecules inside the container can be written as follows:

$$\langle F \rangle = N \cdot \langle f \rangle = \frac{Nm\langle v_x^2 \rangle}{L}. \tag{5.4}$$

The mean square velocity of a gas molecule, $\langle v^2 \rangle$, can be expressed in terms of the mean square values of the velocity components (v_x, v_y, v_z) as

$$\langle v^2 \rangle = \langle v_x^2 \rangle + \langle v_y^2 \rangle + \langle v_z^2 \rangle.$$

On the other hand, taking into account the randomness of the motion of gas molecules, we can assume that the molecules move in the same way along any of the three axes, x, y and z. Thus we have

$$\langle v_x^2 \rangle = \langle v_y^2 \rangle = \langle v_z^2 \rangle = \frac{1}{3}\langle v^2 \rangle. \tag{5.5}$$

The gas pressure on the surface S can be expressed from Eqs. (5.4) and (5.5) as

$$p = \frac{\langle F \rangle}{L^2} = \frac{Nm\langle v^2 \rangle}{3L^3} = \frac{Nm\langle v^2 \rangle}{3V}, \tag{5.6}$$

where $V = L^3$ is the volume of the cube, or equivalently the volume of the gas.

Example 5.1. Evaluate the average kinetic energy of a gas molecule, by using the ideal gas equation (absolute temperature T). Use the Boltzmann constant $k_B \equiv R/N_A$ (R: the gas constant; N_A: Avogadro's number)

Solution

Substituting Eq. (5.6) into Eq. (5.2) (the ideal gas equation), we get

$$\frac{1}{3}Nm\langle v^2\rangle = nRT.$$

Then the average kinetic energy of a gas molecule at temperature T can be expressed in terms of the Boltzmann constant.

$$\frac{1}{2}m\langle v^2\rangle = \frac{3}{2}\frac{nR}{N}T = \frac{3}{2}k_{\mathrm{B}}T, \tag{5.7}$$

where $N = nN_{\mathrm{A}}$ is the total number of molecules, and n is the molar number of the gas. ∎

Example 5.2. Evaluate the root mean square velocity, $\sqrt{\langle v^2\rangle}$, of a molecule of the air at temperature $T = 300\,\mathrm{K}$. Although the air mainly consists of oxygen and nitrogen, for simplicity, you can here assume that the air consists of a single species of gas and that its mass per mole is $30\,\mathrm{g}$. The gas constant is $R = 8.3\,\mathrm{J/K}$.

Solution

From Eq. (5.7), we have

$$\sqrt{\langle v^2\rangle} = \sqrt{\frac{3k_{\mathrm{B}}T}{m}} = \sqrt{\frac{3RT}{N_{\mathrm{A}}m}} = \sqrt{\frac{3\times 8.3\times 300}{30\times 10^{-3}}} = 5.0\times 10^2\,\mathrm{m/s}.$$

∎

5.2.2. *Internal Energy*

Except for the kinetic and potential energies of the matter as a whole, the energy due to the internal motion and displacement of the molecules constituting the matter is called **the internal energy** of the matter.

The internal energy of an ideal gas consisting of monatomic molecules contains only the energy of the translational motion of each molecule. If the molecule of an ideal gas consists of plural atoms, the internal energy contains not only the energy of translational motion but also the energy of rotation and oscillation around the center of mass of the molecule.

An ideal gas of n moles consisting of monatomic molecules (the absolute temperature T) has the internal energy U expressed from (5.7) as follows:

$$U = \frac{3}{2}Nk_{\mathrm{B}}T = \frac{3}{2}nRT, \tag{5.8}$$

where we used the relation, $N = nN_{\mathrm{A}}$.

5.3. The First Law of Thermodynamics

The first law of thermodynamics is the most important law in thermodynamics. It is another expression of energy conservation law where we treat heat as energy.

5.3.1. *Quasi-Static Process*

The process in which the change of the system is sufficiently slow so that its thermal equilibrium is maintained during the change is called the **quasi-static process.** In the quasi-static process, since the system is always kept in thermal equilibrium, the pressure, p, volume, V, and temperature, T, can be determined at any moment. The change of the gas is expressed by $p-V$ graph as shown in Fig. 5.5. The system changed in a quasi-static process can be traced back to the initial state by another quasi-static process. Such a property

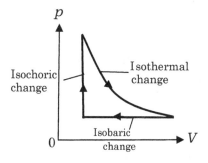

Fig. 5.5.

of being capable of retracing the state change is called **reversible**. The quasi-static process is a **reversible process**.

5.3.2. *The First Law of Thermodynamics*

As shown in Fig. 5.6, we consider a gas contained in a cylinder. Suppose a small quantity of heat, $d'Q$, is added to the gas and a work, $d'W$, is done from outside to the gas in the cylinder. As a result, the internal energy of the gas is increased by dU. Hence, the energy conservation law can be written as

$$dU = d'Q + d'W, \tag{5.9}$$

where the prime on $d'Q$ and $d'W$ are added in order to express that they are freely changeable quantities, unlike the quantity of state such as the internal energy. The law expressed by Eq. (5.9) is called the first law of thermodynamics.

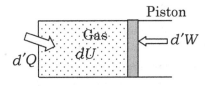

Fig. 5.6.

Let a pressure of the gas in the cylinder be p and the cross section of the cylinder be S, as shown in Fig. 5.7. If the piston moves a distance, dx, the change of the volume of the gas is $dV = Sdx$. So the work done on the gas is expressed as

$$d'W = -pSdx = -pdV,$$

from which we see that when the volume of the gas is changed from V_1 to V_2, the work done on the gas is expressed as

$$W = -\int_{V_1}^{V_2} pdV. \tag{5.10}$$

For ideal gases the intermolecular forces can be ignored. Therefore, the distance among molecules does not affect the internal energy.

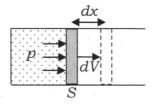

Fig. 5.7.

We have the following property for an ideal gas:

The internal energy of an ideal gas is determined only by its temperature.

If the change of the temperature, ΔT, is the same, the change of the internal energy is identical, however large the volume of the gas changes.

Example 5.3. When the volume of the gas is kept constant, the quantity of heat, C_v, needed to increase the temperature of the gas (1 mole) by 1 K, is called the **molar heat at constant volume**. When the pressure of the gas is kept constant, the quantity of heat, C_p, needed to increase the temperature of the gas (1 mole) by 1 K is called the **molar heat at constant pressure**. By considering the change of an ideal gas (n moles) at constant volume and at constant pressure, show the following equations:

$$dU = nC_v dT, \qquad (5.11)$$

$$C_p - C_v = R. \qquad (5.12)$$

Equation (5.12) is called **Mayer's relation**.

Solution

Suppose a quantity of heat, $d'Q$, is added to the ideal gas of n mol at constant volume, the temperature is increased by dT. Hence, the definition of the molar heat at constant volume yields

$$d'Q = nC_v dT.$$

Because the volume is kept constant, the work done on the gas is zero. Therefore, from the first law of thermodynamics (5.9), we have the following relation:

$$dU = d'Q = nC_v dT.$$

This relation is Eq. (5.11). Because the internal energy is determined only by its temperature, Eq. (5.11) holds true in any processes with changing temperature, not only at constant volume.

Next, a quantity of heat, $d'Q$, is added to the ideal gas at constant pressure, then the temperature is increased by dT. If we denote the pressure of the gas as p and the change of the volume as dV, the first law of thermodynamics is expressed as follows:

$$dU = d'Q - pdV. \tag{5.13}$$

The ideal gas equations before and after the addition of $d'Q$ can, respectively, be written as

$$\begin{cases} pV = nRT \\ p(V + dV) = nR(T + dT). \end{cases}$$

from which we have the relation

$$pdV = nRdT.$$

Therefore, Eq. (5.13) can be expressed as follows:

$$dU = d'Q - nRdT. \tag{5.14}$$

On the other hand, from the definition of the molar heat at constant pressure, we have the relation

$$d'Q = nC_p dT. \tag{5.15}$$

Since Eq. (5.11) holds true even at constant pressure, substitution of Eqs. (5.11) and (5.15) into Eq. (5.14) yields the relation (5.12). ∎

Example 5.4. Considering a small adiabatic change, show that the following relation among the pressure p, volume V and temperature T

holds true in the quasistatic adiabatic process of an ideal gas:

$$pV^\gamma = \text{const.}, \quad TV^{\gamma-1} = \text{const.} \tag{5.16}$$

where $\gamma = C_p/C_v$ is the specific heat ratio and Eq. (5.16) is called **Poisson's equation**.

Solution

Consider a small adiabatic change from p, V, T to $p + dp$, $V + dV$, $T + dT$. The ideal gas equations before and after the change can be written as follows:

$$\begin{cases} pV = nRT \\ (p + dp)(V + dV) = nR(T + dT). \end{cases}$$

By taking the difference between the above two equations and ignoring $dp \cdot dV$, we obtain the relation:

$$pdV + Vdp = nRdT. \tag{5.17}$$

On the other hand, in the adiabatic change $d'Q = 0$, therefore, the first law of thermodynamics is expressed as

$$nC_v dT = -pdV, \tag{5.18}$$

since dU can be described by Eq. (5.11). We eliminate ndT from Eqs. (5.17) and (5.18). Thus the following equation can be derived:

$$\frac{C_v + R}{C_v} \frac{dV}{V} + \frac{dp}{p} = 0 \Rightarrow \gamma \frac{dV}{V} + \frac{dp}{p} = 0,$$

where Mayer's relation (5.12) and the definition for the specific heat ratio, $\gamma = C_p/C_v$, have been used. By integrating both sides of this equation, we get

$$\log pV^\gamma = C \, (C : \text{an integral constant})$$

This is the first equation of (5.16). From this equation and the ideal gas equation, $pV = nRT$, we eliminate p to obtain the second equation of (5.16). ∎

Advanced Problems

Problem 5.5.   Brownian motion

Thermodynamics deals with two types of macroscopic states, equilibrium states and non-equilibrium states. Equilibrium states are the ones in which matter is uniform and static. The equation of states for gas, which gives a definite relation among pressure, volume and temperature, is a typical example related to equilibrium states. In a mixture, the degree of mixing depends on pressure and temperature; normally components mix well at higher temperature and they may be separated at lower temperature. Reverse may be the case for a special mixture.

On the other hand, we observe non-equilibrium phenomena; in everyday life a mixture, which is initially non-uniform, approaches an equilibrium state, which is uniform. Heat flows from the hotter part to the cooler part. This phenomenon is called **heat conduction**. In a mixture, solute flows from the higher concentration part to the lower concentration part. This phenomenon is called **diffusion**. The mechanism behind diffusion is fluctuation, which is caused by collision of solute particles with solvent molecules. Einstein gave a deep insight into the relation between fluctuation and diffusion. Actually if one looks into the equilibrium state in a more detailed way, it is not really static but more dynamic; namely the equilibrium state is attained as the balance between competing directions of evolution, which will be shown below by describing the motion of solute particles in a solvent, as Einstein demonstrated.

Suppose we have insoluble powder particles whose density is higher than that of water. If we put the powder particles into a tube, which is filled with water, then the powder particles start to descend. We may expect that they form sediment at the bottom of the tube, but it is not the case. Rather, we have a gradual change of concentration in the vertical direction; we have lower concentration at a higher position and higher concentration at a lower position. As we will see in Sec. I, the profile of powder concentration is determined by the balance between two forces acting on the particles. Einstein further thought that the profile is determined by the condition for the balance between a flow of descending particles due to gravity and a flow of ascending particles due to diffusion from a high concentration position to a low concentration position.

So there are two points of view for the same phenomenon; one is the static view of the balance of forces and the other is the dynamical view of the balance of competing flows. Combining these two views, Einstein could find a relation, called **Einstein's relation**.

In sections I to III below, we will derive the diffusion coefficient Einstein found in 1905. Let g be the magnitude of gravity acceleration.

I Concentration of powder particles and osmotic pressure

Powder particles scattered in the water with sufficiently low concentration are known to behave like ideal gas molecules. If we denote the gas constant with R, we have the equation of state, $pV = RT$, for one mole of gas molecules, where p is pressure, V is volume and T is absolute temperature. Similarly, if we denote the osmotic pressure of N powder particles within volume V with p and Avogadro's number with N_A', we have $pV = \frac{N}{N_A} RT$. Here T is the absolute temperature of water.

(1) Express the osmotic pressure p in terms of the concentration of powder particles $n = \frac{N}{V}$, the Boltzmann constant $k_B = \frac{R}{N_A}$ and T.

The concentration of the powder particles and its osmotic pressure depends on the altitude. We denote the concentration and the osmotic pressure at altitude h with n and p, respectively and the concentration and the osmotic pressure at altitude $h + \Delta h$ with $n+\Delta n$ and $p+\Delta p$, respectively. We also assume that the temperature is uniform regardless of the altitude.

(2) From the equation derived in part (1), express the difference of osmotic pressure, Δp, in terms of the difference of concentration, Δn.

Let us consider a rectangular parallelepiped of cross section area A and height Δh (with the lower cross section at h and with the

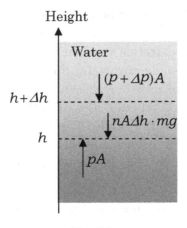

Fig. 5.8.

higher cross section at $h+\Delta h$) within the water and study the balance of forces acting on the powder particles within the parallelepiped. Let us denote the mass of each particle with m. There are three forces, gravity force, force due to osmotic pressure from the upper cross section and force due to osmotic pressure from the lower cross section. We neglect buoyancy force by assuming the density of a particle to be sufficiently larger than that of water.

Since Δh is small, we may assume that the concentration n inside the parallelepiped is constant. Then the total number of particles within the parallelepiped is $nA\Delta h$. The gravity force on the powder particles is now $nA\Delta h \cdot mg$. The force due to the osmotic pressure on the upper cross section is $(p + \Delta p)A$, while the force due to the osmotic pressure on the lower cross section is pA.

(3) From the balance of these three forces, express Δp in terms of Δh.

(4) Using the results of (2) and (3), express the concentration gradient $\frac{\Delta n}{\Delta h}$ in terms of g, k_B, m, n and T.

The static view above does not give any insight into the dynamics of powder particles. In Secs. II and III below, we will give the way to measure the dynamics.

II Mobility of particles

We set a long cylindrical tube filled with water in the vertical direction. We add a particle of mass m in the tube. Then the particle moves downwards under the action of friction due to viscosity of water. Stokes' law is known for the friction F, which is proportional to the velocity u of the particle, namely,

$$F = Cau,$$

where C is a constant determined by the viscosity of the water and a is the radius of the particle. The particle will attain the ultimate velocity, which is proportional to the gravity force mg, namely the ultimate velocity is given by $u = Bmg$, where B is called **mobility**.

(5) Express the mobility B in terms of a and C.

From this, the down flow J of particles determined by the gravity and the viscosity, which is the number of particles that cross a horizontal section per unit time and per unit area, is given by $J = nu = nBmg$.

III Diffusion coefficient and Einstein's relation

Let us consider the case that the concentration of particles depends on the position x.

The concentration at x is denoted by n and the concentration at $x + \Delta x$ is denoted by $n + \Delta n$. Hence, the concentration gradient can be expressed by $\frac{\Delta n}{\Delta x}$. As shown in Fig. 5.9, the diffusion flow $J(x)$, which is the number of particles crossing a section at x from the left to the right per unit time per unit area, is proportional to the magnitude of the concentration gradient in the direction opposite to the concentration gradient, namely

$$J(x) = -D\frac{\Delta n}{\Delta x}, \qquad (5.19)$$

where D is called the **diffusion coefficient**.

As explained in Sec. II, the down flow due to the gravity and the up flow due to the diffusion will be balanced, and its balance

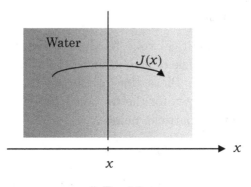

Fig. 5.9.

determines the equilibrium profile of the concentration obtained in Sec. I.

(6) Express the relation between D and B in terms of k_B and T. This relation is called **Einstein's Relation**.

(7) Determine the diffusion coefficient $D\,[\mathrm{m^2/s}]$ for particles of radius $1.0\,\mu\mathrm{m} = 1.0 \times 10^{-6}\,\mathrm{m}$ in the water of 20°C with two significant figures. Here $k_B = 1.38 \times 10^{-23}\,\mathrm{J/K}$ and the constant C determined by the viscosity of the water is given by $C = 2.00 \times 10^{-2}\,\mathrm{Pa \cdot s}$ (20°C).

The diffusion coefficient D determined here does not depend on the gravity force acting on the particle.

IV Particle colliding with water molecules

We now put water and particles in a horizontal vessel so that the concentration distribution is not influenced by the gravity, and then observe the diffusion of particles. We take the horizontal direction as the x-axis, as shown in Fig. 5.10.

As we mentioned in Sec. III, let us, furthermore, consider why the diffusion flow is proportional to the magnitude of the concentration gradient. Actually, particles change their directions of motion when they collide with water molecules. Let us denote the average velocity of particles with $\langle v \rangle$. They change the directions of motion in a time

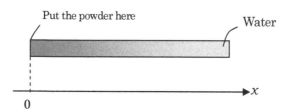

Fig. 5.10.

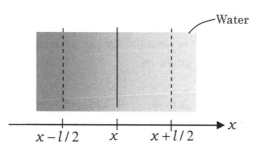

Fig. 5.11.

interval t_m, called **mean free time**. The length $l = \langle v \rangle t_m$ is called **mean free path**, within which particles can move without collisions.

The number of particles flowing from the left region to the right region per unit time per unit area of the cross section at x is proportional to the average velocity $\langle v \rangle$ and the concentration $n(x - l/2)$ at position $x - l/2$, as shown in Fig. 5.11. Furthermore, the particles have velocities of various directions. For simplicity, we assume that the velocities are distributed equally in positive and negative directions of the x-, y- and z-axes. Then we may assume that one sixth of particles move in the positive direction of x-axis. Therefore, the number of particles flowing from the left region to the right region per unit time per unit cross section at x is $\frac{1}{6}\langle v \rangle n(x - l/2)$. Similarly the number of particles flowing from the right region to the left region per unit time per unit cross section at x is given by $\frac{1}{6}\langle v \rangle n(x + l/2)$.

The diffusion flow $J(x)$ is considered the difference between the two flows.

(8) When l is sufficiently small, we may have

$$n(x - l/2) = n(x) - \frac{l}{2}\frac{\Delta n}{\Delta x}, \quad n(x + l/2) = n(x) + \frac{l}{2}\frac{\Delta n}{\Delta x}.$$

Express D in terms of l and $\langle v \rangle$.

V Behavior of a particle in the diffusion

Let us consider how a particle behaves in the diffusion. During a time interval t, the particle changes its direction $N = \frac{t}{t_m}$ times. The displacement of the particle between the i-th change of its direction and the $(i+1)$-th change is denoted by Δx_i. Then the position x at the $(N+1)$-th change is given by

$$x = \Delta x_1 + \Delta x_2 + \cdots + \Delta x_N = \sum_{i=1}^{N} \Delta x_i. \tag{5.20}$$

Each displacement is irregular and so the average vanishes; namely $\langle \Delta x_i \rangle = 0$. Therefore the average of the displacement at time t vanishes; namely $\langle x \rangle = 0$. To understand the diffusion, let us consider the average of the square of the displacement. From Eq. (5.20), we have

$$\langle x^2 \rangle = \sum_{i=1}^{N} \langle (\Delta x_i)^2 \rangle + \sum_{\substack{i,j \\ i \neq j}} \langle \Delta x_i \Delta x_j \rangle. \tag{5.21}$$

If we consider that the displacement of each step is irregular and independent from the displacements of other steps, we may assume

$$\langle (\Delta x_1)^2 \rangle = \langle (\Delta x_2)^2 \rangle = \cdots = \langle (\Delta x_N)^2 \rangle \neq 0,$$

$$\langle \Delta x_i \Delta x_j \rangle = \langle \Delta x_i \rangle \langle \Delta x_j \rangle = 0 \ (i \neq j).$$

(9) Using the fact that the number of changing velocity during the time interval t is written as $N = \frac{t}{t_m}$, express the mean square displacement $\sqrt{\langle x^2 \rangle}$ at time t in terms of the diffusion coefficient D and t.

Here, assume the displacement Δx at each time is independent of the one at another time. If you denote the x component of the velocity with v_x, you may write $\Delta x = v_x t_\text{m}$. Because you can neglect the influence of gravity on particles, the motion in each of the three directions, x, y and z, are independent of each other. If you write the y and z components of the velocity as v_y and v_z, respectively, you may have $\langle v_x^2 \rangle = \langle v_y^2 \rangle = \langle v_z^2 \rangle = \frac{1}{3}\langle v^2 \rangle$ from $\langle v_x^2 \rangle + \langle v_y^2 \rangle + \langle v_z^2 \rangle = \langle v^2 \rangle$. You may also use $\langle v \rangle \approx \sqrt{\langle v^2 \rangle}$.

(10) From the discussion above, you can see that particles in the vessel will spread over the whole region of the vessel. Obtain the time t for the particles to spread over the whole vessel with two significant figures by assuming the length of the vessel to be 10 cm and the radius of the particle to be $1.0\,\mu\text{m} = 1.0 \times 10^{-6}$ m. Suppose the temperature of the water is 20°C. From the result, discuss whether it is realistic to wait for powder particles to spread over the whole vessel without stirring the water.

(the 2nd Challenge)

Solution

(1) Using the concentration n of the particles and the Boltzmann constant k_B, we have the ideal gas equation:

$$p = \frac{N}{V}\frac{R}{N_\text{A}}T = nk_\text{B}T.$$

Namely, $\underline{p = nk_\text{B}T}$.

(2) The ideal gas equations at altitudes h and $h + \Delta h$ are

$$p = k_\text{B}Tn, \quad p + \Delta p = k_\text{B}T(n + \Delta n),$$

respectively. Then we have

$$\Delta p = \underline{k_\text{B}T\Delta n}.$$

(3) The balance of the forces on particles in the parallelepiped is

$$(p + \Delta p)A + nmg \cdot A\Delta h = pA,$$

which yields $\underline{\Delta p = -nmg \cdot \Delta h}$.

(4) Using the results of parts (2) and (3), we have

$$\frac{\Delta n}{\Delta h} = -\frac{nmg}{k_B T}.$$

(5) When the particle attains its ultimate velocity, the gravity force and the friction are balanced. The balance is expressed by $mg = Cau$, which yields

$$mg = Cau = Ca \cdot Bmg, \quad \therefore \ B = \frac{1}{Ca} \qquad (5.22)$$

(6) Since the concentration of particles is larger in the lower part of the tube, the diffusion flow $-D\frac{\Delta n}{\Delta h}$ is in the ascending direction. The balance of the ascending flow and the descending flow, $J = nBmg$, is given by the usage of the result of part (4) as

$$D\frac{nmg}{k_B T} = nBmg \quad \therefore \quad D = k_B T B. \qquad (5.23)$$

(7) From Eqs. (5.22) and (5.23), we obtain

$$D = \frac{1.38 \times 10^{-23} \times 293}{2.00 \times 10^{-2} \times 1.0 \times 10^{-6}} = 2.0 \times 10^{-13} \, \mathrm{m^2/s}.$$

(8) $J(x)$ is obtained as the difference between the number of particles flowing from the left to the right and that from the right to the left, per unit time per unit cross section at x. Therefore we have

$$J(x) = \frac{1}{6}\langle v \rangle n(x - l/2) - \frac{1}{6}\langle v \rangle n(x + l/2)$$

$$= \frac{1}{6}\langle v \rangle \left\{ \left(n(x) - \frac{l}{2} \cdot \frac{\Delta n}{\Delta x} \right) - \left(n(x) + \frac{l}{2} \cdot \frac{\Delta n}{\Delta x} \right) \right\}$$

$$= -\frac{1}{6}\langle v \rangle l \frac{\Delta n}{\Delta x}.$$

By comparing this with Eq. (5.19), we obtain

$$D = \frac{1}{6}\langle v \rangle l.$$

(9) In Eq. (5.21), using $\Delta x = v_x t_m$ and $\langle v_x^2 \rangle = \langle v_y^2 \rangle = \langle v_z^2 \rangle = \frac{1}{3}\langle v^2 \rangle$, we have

$$\langle x^2 \rangle = N\langle (\Delta x)^2 \rangle = N\langle v_x^2 \rangle t_m^2 = \frac{1}{3}\frac{t}{t_m}\langle v^2 \rangle t_m^2.$$

Further, using $l = \langle v \rangle t_m \approx \sqrt{\langle v^2 \rangle} t_m$ and $6D = \langle v \rangle l \approx \sqrt{\langle v^2 \rangle} l$, we have

$$\langle x^2 \rangle = 2Dt \quad \therefore \quad \sqrt{\langle x^2 \rangle} = \sqrt{2Dt}.$$

(10) From the result of part (9), we have

$$t = \frac{\langle x^2 \rangle}{2D}.$$

When $\sqrt{\langle x^2 \rangle}$ becomes around $10\,\mathrm{cm}$, we can say that the particles have spread over the whole vessel. We put $\sqrt{\langle x^2 \rangle} = 10\,\mathrm{cm} = 0.10\,\mathrm{m}$ and $D = 2.0 \times 10^{-13}\,\mathrm{m^2/s}$, we obtain

$$t = \frac{0.10^2}{2 \times 2.0 \times 10^{-13}} = 2.5 \times 10^{10}\,\mathrm{s} \approx 790 \text{ years},$$

which implies that the particles will never spread over the whole vessel unless we stir the water. ∎

Problem 5.6.   Thermal conduction

There are an enormously large number (roughly $1/10$ of Avogadro's number $N_A = 6.02 \times 10^{23}$) of gas molecules inside an empty large-size plastic bottle. The molecules that constitute the gas are moving almost freely, colliding occasionally with one another and interacting among themselves through weak forces. In a collection of such a large number of molecules its physical behavior can be expressed by macroscopic quantities, which are obtained by statistically averaging the behaviors of individual molecules. Temperature, pressure, flow of gas, etc. are such macroscopic physical quantities that are observable to us.

In the present study, we will consider the kinetic theory of gases that interprets the macroscopic properties of a gas in terms of the microscopic behaviors of the constituent individual molecules. Let's try to explain the mechanism of a thermos bottle through a step-by-step approach based on the kinetic theory of gases.

In the following, the gravitational force acting on a gas molecule is neglected because the kinetic energy of the molecule is extremely large as compared with the change in the gravitational potential energy of the molecule.

I In a gas where temperature and pressure are uniform, the Boyle–Charles law holds almost exactly in a wide range of gas parameters. When a gas of pressure p occupies volume V at absolute temperature T, the mathematical expression for this law is

$$\frac{pV}{T} = \text{const.} \tag{5.24}$$

where the constant on the right-hand side is proportional to the amount of the gas. There is another gas law that under the same condition of temperature and pressure, all kinds of gases contain the same number of molecules if their volumes are the same. It is known that one mole of gas in the standard condition for temperature and pressure (0°C, 1 atm) occupies the volume of $2.24 \times 10^{-2} \, \text{m}^3$ and the gas contains molecules of Avogadro's number, independently from the kind of the gas.

(1) Find the volume allotted to one molecule of a gas under the standard condition for temperature and pressure.

(2) Suppose a gaseous molecule is a rigid sphere whose radius is $r = 1.0 \times 10^{-10}$ m. What is the ratio of the volume allotted to one molecule in the standard condition of the gas to a molecule's own volume?

II Suppose N gas molecules of mass m are confined in a box of volume V. A distance between two face-to-face end walls of the box, which are perpendicular to the x-axis, is L, and the area of each wall is S, so that the volume of the box is given by $V = LS$. Let molecules be moving in the box along the x-axis without colliding among themselves and let the velocity of i-th molecule be v_i. The molecules are supposed to collide elastically with the wall, which is assumed to be fixed. Each gas molecule hits at and bounces off the wall, travels to the opposite direction with the same speed, collides with the opposite wall, and moves back and forth with the same speed.

When a molecule collides with the wall, it exerts an impulsive force on the wall. Although the forces exerted by individual collisions are impulsive, the forces exerted on the wall are averaged to give an almost constant force, because the number of collisions by all the molecules in a unit time is enormously large. The pressure of the gas is the average force exerted by the molecules per unit area on the wall.

(3) Find the expression for the time-averaged force f_i, or the impulse per unit time, exerted on the wall by the i-th molecule moving with velocity v_i. Also find the total force F, which is the sum of the time-averaged forces exerted by all molecules.

(4) Show that the pressure of gas is proportional to the number of molecules per unit volume, $\frac{N}{V}$, by using the result of part (3).

(5) Derive the Boyle–Charles law (5.24), by using the above results and the fact that the average kinetic energy of a molecule is proportional to the absolute temperature of the gas, T.

III In Sec. II we neglected the collisions among gas molecules. In this section, we treat molecules based on a more realistic model in which they move randomly colliding with one another and changing their speeds and directions in chaotic fashion.

(6) In this model of the molecular motion, the distribution of molecular velocities is supposed to be identical in every direction and therefore the average values of the squares of the velocity components, v_x, v_y and v_z, are equal to each other: $\langle v_x^2 \rangle = \langle v_y^2 \rangle = \langle v_z^2 \rangle$. Using this relation and the ideal gas equation, express the average kinetic energy of a molecule in terms of the absolute temperature T and the Boltzmann constant $k_B = \frac{R}{N_A}$ (R is the gas constant).

We now introduce the concept of the mean free path. As its name implies, it is the average distance traversed by a molecule between two successive collisions. Let the mean free path be l.

(7) Suppose the molecules are spheres of radius r, then we expect that on average there is a single molecule within a cylinder of cross-sectional radius $2r$ and axis length l (see Fig. 5.12). From such a consideration, find the expression for l in terms of r and N/V, being the number of molecules per unit volume.

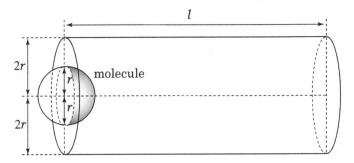

Fig. 5.12.

(8) Using the ideal gas equation, calculate the mean free path of helium gas at $T = 273\,\mathrm{K}$ and 1 atm ($=1.0 \times 10^5\,\mathrm{Pa}$) with two significant figures. Use the following values: the molar mass of helium gas $4.0 \times 10^{-3}\,\mathrm{kg/mol}$, the radius of helium atom $r = 1.0 \times 10^{-10}\,\mathrm{m}$, and the Boltzmann constant $k_\mathrm{B} = 1.4 \times 10^{-23}\,\mathrm{J/K}$.

Based on the above argument, let us discuss the heat conduction due to a temperature gradient within a gas. Heat flows from a region at high temperature to a region at low temperature. Heat flow due to a temperature gradient is in close analogy with electric current in a conductor due to an electric potential difference. Let us introduce the coordinate shown in Fig. 5.13 and denote the absolute temperature at the position x by $T(x)$, thus the law of heat conduction is given by the equation

$$Q = -k\frac{\Delta T}{\Delta x},\qquad(5.25)$$

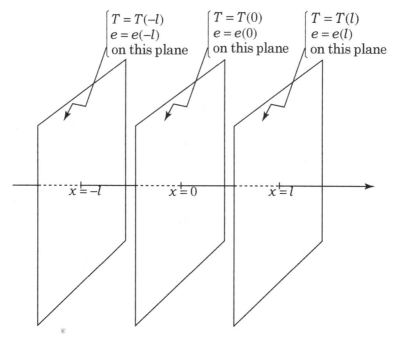

$$\begin{bmatrix} T = T(-l) \\ e = e(-l) \\ \text{on this plane} \end{bmatrix} \quad \begin{bmatrix} T = T(0) \\ e = e(0) \\ \text{on this plane} \end{bmatrix} \quad \begin{bmatrix} T = T(l) \\ e = e(l) \\ \text{on this plane} \end{bmatrix}$$

Fig. 5.13.

where Q is the heat flux, i.e., the quantity of heat transferred per unit time and unit cross-sectional area, Δx is an infinitesimal distance along the x-axis, and ΔT is the temperature difference in the distance Δx, so that $\frac{\Delta T}{\Delta x}$ is the temperature gradient. The proportional coefficient k is called the **thermal conductivity**, whose unit is $W/(m{\cdot}K)$.

As obviously shown in Eq. (5.25), the more the thermal conductivity is and the greater the temperature gradient is, the greater the quantity of heat flows. The negative sign on the right-hand side of Eq. (5.25) means that when $k > 0$ the quantity of heat is transferred from a high-temperature region to a low-temperature region.

The heat conduction in a gas is caused by the energy transferred from molecules of high energy to those of low energy through their collisions, so it can be treated based on the model of gas molecular motion in which collisions are taken into account.

We now derive the formula for thermal conductivity k using the following simplified model. Let us consider the quantity of heat transferred from the region of $x < 0$ to the region of $x > 0$ through the plane $x = 0$ at temperature $T(0)$ (see Fig. 5.13).

Assume the molecule that passes through the plane $x = 0$ has energy $e(\pm l)$, which is the average kinetic energy of the molecule at the last collision on the plane $x = \pm l$ separating the mean free path from the plane $x = 0$. Let the number of molecules that pass through a unit area of the plane $x = 0$ per unit time from the region of $x < 0$ to the region of $x > 0$ be f_+, and the number of those from the region of $x > 0$ to the region of $x < 0$ be f_-. As both of these numbers are proportional to the product of the average molecular velocity $v = \sqrt{\langle v^2 \rangle}$ on the plane $x = 0$ and the number of molecules in a unit volume N/V, f_+ and f_- are equal to each other. Therefore we denote them by f, which is given by

$$f = f_+ = f_- = \frac{\alpha N v}{V},$$

where α is a proportional constant.

The quantity of heat Q transferred through a unit area of the plane $x = 0$ per unit time from the region of $x < 0$ to the region of $x > 0$ is given by $Q = Q_+ - Q_-$, where $Q_+ = f \cdot e(-l)$ is the kinetic energy of the molecules transferred from the region of $x < 0$, and $Q_- = f \cdot e(+l)$ is that from the region $x > 0$.

(9) Using the relation

$$e(\pm l) = e(0) \pm l \left. \frac{\Delta e(x)}{\Delta x} \right|_{x=0} \qquad \text{(double-sign in same order)}$$

and the result of part (6), find the expression for the thermal conductivity k of gases. Here $\left. \frac{\Delta e}{\Delta x} \right|_{x=0}$ is the rate of change of the average kinetic energy e with respect to x on the plane $x = 0$. If necessary, you can use the relation

$$\left. \frac{\Delta e}{\Delta x} \right|_{x=0} = \left. \frac{\Delta e}{\Delta T} \right|_{x=0} \cdot \left. \frac{\Delta T}{\Delta x} \right|_{x=0}.$$

Then calculate the thermal conductivity k of helium gas at $T = 273\,\mathrm{K}$ with two significant figures, letting $\alpha = 1$.

(10) We now apply the formulae derived for the heat conduction to a thermos bottle. The thermos bottle is a double-walled container, with a near vacuum between the walls to prevent the heat conduction.

Suppose the two walls are separated $d = 1.0\,\mathrm{cm}$ from each other and helium gas of 1 atm is filled between the walls. By lowering the helium gas pressure, we are planning to reduce an outflow of heat to about $\frac{1}{100}$. For that purpose, to what extent should you lower the pressure of the helium gas? Find the necessary pressure for the case where the gas temperature is $T = 273\,\mathrm{K}$. Neglect thermal radiation from the walls.

Hint When the mean free path l becomes longer than the distance d between the walls, the molecules carry the energy obtained on one wall straight to the opposite wall. In this case, we should use the distance d rather than the mean free path l.

(the 2nd Challenge)

Solution

(1) Dividing the volume of a mole of gas by the number of molecules in a mole of gas results in

$$\frac{2.24 \times 10^{-2}}{6.02 \times 10^{23}} = \underline{3.72 \times 10^{-26}\,\mathrm{m}^3}.$$

(2) The volume of a molecule is $\frac{4\pi r^3}{3} = 4.19 \times 10^{-30}$. Dividing the previous result by the volume of a molecule, we obtain

$$\frac{3.72 \times 10^{-26}}{4.19 \times 10^{-30}} = \underline{8.88 \times 10^3\ (\text{times})}.$$

Therefore the volume allotted to one molecule is about ten thousand times the volume of a molecule itself.

(3) The impulse that i-th molecule exerts on the wall by a collision is $2mv_i$, which is equal to the change of its momentum, and the

time for this molecule to travel a round trip between the two walls is $\frac{2L}{v_i}$. Therefore the molecule collides $\frac{v_i}{2L}$ times per unit time with one wall. The average force of the molecule exerting on the wall, f_i, is equal to the impulse per unit time, that is

$$f_i = 2mv_i \times \frac{v_i}{2L} = \underline{\frac{mv_i^2}{L}}.$$

The total force F summed over all the molecules is

$$F = \underline{\frac{1}{L} \sum_{i=1}^{N} mv_i^2}.$$

(4) Because the pressure of the gas is $p = \frac{F}{S}$,

$$p = \frac{1}{LS} \sum_{i=1}^{N} mv_i^2 = \frac{1}{V} \sum_{i=1}^{N} mv_i^2.$$

By using the relation $\frac{1}{N} \sum_{i=1}^{N} v_i^2 = \langle v^2 \rangle$ for the mean square velocity of molecules, we have the equation

$$p = \frac{N}{V} m \langle v^2 \rangle. \tag{5.26}$$

Namely, p is proportional to N/V.

(5) The average kinetic energy per molecule is given by

$$\frac{1}{N} \sum_{i=1}^{N} \left(\frac{1}{2} mv_i^2 \right) = \frac{1}{2} m \langle v^2 \rangle.$$

Combining this relation with Eq. (5.26) of part (4), we obtain

$$p = 2 \frac{N}{V} \frac{1}{2} m \langle v^2 \rangle \propto \frac{NT}{V}.$$

Hence it follows that $\frac{pV}{T} = \text{const.}$

(6) In the three dimensional case, the average kinetic energy per molecule is given by

$$e = \frac{1}{2}m(\langle v_x^2 \rangle + \langle v_y^2 \rangle + \langle v_z^2 \rangle) = \frac{3}{2}m\langle v_x^2 \rangle.$$

By using the one dimensional result (5.26) with $\langle v^2 \rangle$ replaced by $\langle v_x^2 \rangle$, we obtain

$$e = \frac{3}{2}m\langle v_x^2 \rangle = \frac{3pV}{2N}.$$

Substitution of the ideal gas equation, $pV = \frac{N}{N_A}RT$, leads to the relation

$$e = \frac{3RT}{2N_A} = \frac{3}{2}k_B T.$$

(7) Because on average there is one molecule in the cylinder of length l and cross-sectional radius $2r$, it follows that

$$N\pi(2r)^2 l = V \quad \therefore \quad l = \frac{1}{4\pi r^2}\frac{1}{N/V}.$$

(8) From the ideal gas equation it follows that

$$pV = Nk_B T \quad \therefore \quad \frac{N}{V} = \frac{p}{k_B T}.$$

Substitution of this relation into the result of part (7) leads to the following equation:

$$l = \frac{k_B T/p}{4\pi r^2}. \tag{5.27}$$

By using the given numerical values, we obtain

$$l = 3.0 \times 10^{-7}\,\text{m}.$$

(9) The given relations yield

$$Q_+ = fe(-l) = \alpha \frac{N}{V} v \left\{ e(0) - l \left. \frac{\Delta e(x)}{\Delta x} \right|_{x=0} \right\},$$

$$Q_- = fe(+l) = \alpha \frac{N}{V} v \left\{ e(0) + l \left. \frac{\Delta e(x)}{\Delta x} \right|_{x=0} \right\},$$

from which we obtain the following equation:

$$Q = Q_+ - Q_- = -2\alpha \frac{N}{V} vl \left. \frac{\Delta e(x)}{\Delta x} \right|_{x=0}.$$

From the relation $e = \frac{3}{2} k_B T$, we can write

$$\left. \frac{\Delta e(x)}{\Delta x} \right|_{x=0} = \left. \frac{\Delta e}{\Delta T} \right|_{x=0} \cdot \left. \frac{\Delta T}{\Delta x} \right|_{x=0} = \frac{3}{2} k_B \left. \frac{\Delta T}{\Delta x} \right|_{x=0}.$$

Combining the last two equations, we obtain

$$Q = -3\alpha k_B \frac{N}{V} lv \left. \frac{\Delta T}{\Delta x} \right|_{x=0}. \tag{5.28}$$

Substitution of the relations, $\frac{N}{V} l = \frac{1}{4\pi r^2}$ and $v = \sqrt{\frac{3k_B T}{m}}$, into the above equation leads to

$$Q = \frac{3\alpha k_B}{4\pi r^2} \sqrt{\frac{3k_B T}{m}} \left. \frac{\Delta T}{\Delta x} \right|_{x=0}.$$

Hence the following expression for k is obtained.

$$k = \frac{3\alpha k_B}{4\pi r^2} \sqrt{\frac{3k_B T}{m}}.$$

By using $\alpha = 1$, $T = 273$ K and the data for the helium atom, we obtain

$$k = 0.44 \, \text{W}/(\text{m} \cdot \text{K}).$$

(10) According to the result of part (9), the thermal conductivity of a gas depends only on temperature T, but it does not depend on pressure at constant temperature.

In the following, we consider the temperature of the gas to be constant. The result of part (8) states that the mean free path is inversely proportional to pressure and hence it increases as pressure decreases. When mean free path l becomes longer than the distance between the walls $d = 1.0 \times 10^{-2}$ m, l in Eq. (5.28) is replaced by the distance d as we have pointed out in the hint. Then the thermal conductivity that is proportional to $\frac{N}{V}d$ decreases as the number density $\frac{N}{V}$, i.e., the pressure p, decreases.

From Eq. (5.27) in part (8), the mean free path at pressure of 1 atm ($p_0 = 1.0 \times 10^5$ Pa) is $l = 3.0 \times 10^{-7}$ m and the pressure at which mean free path l is equal to $d = 1.0 \times 10^{-2}$ m is

$$p = \frac{l}{d}p_0 = 3.0\,\text{Pa}.$$

When the pressure is lowered less than 3.0 Pa, l is replaced by d and the thermal conductivity of the gas decreases in proportion to the pressure. Therefore if the pressure is lowered to $1/100$ of the above value, the thermal conductivity is also reduced to $1/100$.

The above consideration shows that in order to reduce the outflow of heat to $1/100$, the pressure of the gas should be lowered to

$$\underline{3.0 \times 10^{-2}\text{Pa}.} \qquad \blacksquare$$

Chapter 6

Modern Physics

Elementary Problems

Problem 6.1.   Tests of general relativity

Because an atomic nucleus consists of several elementary particles, it has various internal states with intrinsic energies. Any nucleus has internal energy in addition to the mechanical energy associated with the motion in a space.

When the state of an atomic nucleus changes from a higher energy state to a lower energy state, the nucleus emits an electromagnetic wave of high frequency (very short wavelength), which is called a **gamma ray**. An electromagnetic wave such as a gamma ray exhibits wave-like and particle-like properties simultaneously. Accordingly, we consider that a gamma ray consists of small packages of energy called **photons**. The energy and momentum of a photon depend on the frequency of the corresponding electromagnetic wave. The energy of the photon with frequency ν is given by $h\nu$ (h is called **Planck's constant**), and its momentum is $\frac{h\nu}{c}$ where c is the speed of light in vacuum. The direction of the momentum is the same as that of the propagation. The photon behaves as a massless particle with the momentum but it differs from a material particle whose momentum is defined by the product of the mass m and the velocity v.

We consider for a while that an atomic nucleus is fixed at a point in a space. As shown in Fig. 6.1, when the nucleus makes a transition from an energy state E_H to an energy state E_L, it emits a gamma ray of frequency ν_0, with the energy given by

$$h\nu_0 = E_H - E_L. \tag{6.1}$$

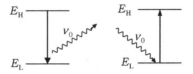

Fig. 6.1.

Inversely, a fixed nucleus with the energy E_L can absorb the gamma ray with the same energy as the energy difference between the two states, and then it can be excited to the energy state E_H. This process is called the **resonance absorption**.

In a general case where the nucleus can move freely in a space, the frequency of the emitted gamma ray does not satisfy Eq. (6.1) and the gamma ray cannot be absorbed by another nucleus in the energy state E_L.

(1) When a free atomic nucleus emits a gamma ray, it recoils in the direction opposite to the direction of the propagation of the gamma ray as required by the law of momentum conservation (see Fig. 6.2). Suppose a nucleus at rest emits a gamma ray with frequency ν. Then, find an expression for the speed of the recoiling nucleus, denoting the mass of the nucleus by M.

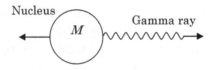

Fig. 6.2.

(2) Find an expression for the kinetic energy K of the recoiling nucleus. The kinetic energy K is called the **recoil energy**.

(3) When the nuclear energy is reduced from E_H to E_L, the energy $h\nu$ of the emitted photon is equal to $h\nu_0 - K$ and the frequency ν of the gamma ray is slightly less than ν_0. Let the energy of the emitted photon be E. Calculate the value of K/E, taking into account that $h\nu \approx h\nu_0$ and $\frac{h\nu}{Mc^2} = \frac{1}{4 \times 10^6}$.

Because the photon energy is decreased by this small recoil energy, it is not sufficient to excite a nucleus with the energy E_L to the energy state E_H. Thus, it is difficult to observe the resonance absorption of the gamma ray by freely moving nuclei. In a solid, however, atomic nuclei are rigidly bound to each other, and hence the recoil energy is derived by replacing M with the total mass of the solid. This recoil energy is practically zero and the gamma ray can be absorbed by another atomic nucleus of the same element in the solid. This phenomenon is called the **Mössbauer effect**.

When a radiation source or an absorber is moving, the resonance absorption is influenced by the **Doppler effect**. The Doppler effect is well known as the change in frequency of observed sound when the source and the observer are moving relative to each other. Similar effect occurs for light as well.

We, first, summarize the Doppler effect for sound.

When a source of sound with frequency ν_0 approaches a stationary observer at velocity v, the observed frequency ν can be written in terms of the speed of sound V as follows:

$$\nu = \frac{V}{V - v}\nu_0 = \frac{1}{1 - (v/V)}\nu_0. \tag{6.2}$$

When an observer moves with velocity u toward a stationary source, the sound velocity relative to the observer will change to $V + u$ and the observed frequency ν can be evaluated as follows:

$$\nu = \frac{V + u}{V}\nu_0 = \left(1 + \frac{u}{V}\right)\nu_0.$$

Next, let us consider the Doppler effect of electromagnetic waves.

When a source of light with the frequency ν_0 approaches a stationary observer at velocity v, which is much smaller than the speed of light c, the observed frequency ν is given by the same formula as the Doppler effect for the sound emitted from a moving source:

$$\nu = \frac{1}{1 - (v/c)}\nu_0. \tag{6.3}$$

When the speed of the light source, v, is close to the speed of light, c, we cannot neglect the effect of **time dilation** predicted by the theory of special relativity. The time dilation in that theory means that, the time interval T between two events measured in the frame where the observer is at rest is given by

$$T = \frac{T_0}{\sqrt{1 - (v/c)^2}},$$

in terms of the time interval T_0 in the frame where the light source is at rest. This relation can be applied to the period of oscillation of the light and, accordingly, ν_0 in Eq. (6.3) should be replaced by $\sqrt{1 - (v/c)^2}\nu_0$. Hence, by multiplying $\sqrt{1 - (v/c)^2}$ to the right-hand side of Eq. (6.3), we can obtain a frequency ν measured by the observer as follows:

$$\nu = \frac{\sqrt{1 - (v/c)^2}}{(1 - v/c)}\nu_0 = \sqrt{\frac{1 + (v/c)}{1 - (v/c)}}\nu_0. \tag{6.4}$$

The frequency of the gamma ray is also shifted by the gravitational force. When we observe photons emitted from a high place above the ground, we observe that the energy and the frequency of the photons increase. This phenomenon for massless photons is not comprehended by the law of universal gravitation but by the theory of general relativity.

The time interval between two events measured on the ground is shorter than that between two events measured at a higher place. It shows the time dilation due to the gravity.

Let us evaluate an effect in the theory of general relativity.

The inertial force acts on every object in a reference frame moving with acceleration. A motion of a free-fall body under the influence of gravitational and inertial forces is equivalent to that falling in the inertial reference frame which is not under the influence of gravity.

(4) Supposing that you are free falling, consider the case that you detect a flash of light with frequency ν_0 emitted from a point of height H above the ground. It takes time $t = \frac{H}{c}$ for the flash to travel from this point to the ground. During the time $t = \frac{H}{c}$, you gain the speed $v = \frac{gH}{c}$ where g is the acceleration of gravity. You can, then, consider a stationary observer on the ground to

be rising relative to you with the speed $v = \frac{gH}{c}$. Thus, you may interpret the increase of frequency not by the gravity but by the Doppler effect. Assuming $v/c \ll 1$, find an expression for the ratio $\frac{\Delta\nu}{\nu_0}$ of an increase $\Delta\nu$ of frequency to the original frequency ν_0 by using the approximation formula, $\frac{1}{1-x} \approx 1+x$ when $x \ll 1$.

In 1960, Robert Pound and his graduate student Glen A. Rebka Jr. found the frequency shift caused by the gravity, using the height difference of $22\,\mathrm{m}$. They oscillated the radiation source so that the Doppler shift by the motion of the source compensates the frequency shift due to the gravity, and observed the resonance absorption. By this experiment the theory of general relativity was verified.

(the 1st Challenge)

Solution

(1) Denoting the recoil speed of the nucleus by v and using the conservation law of momentum, we have

$$Mv - \frac{h\nu}{c} = 0 \quad \therefore \quad v = \frac{h\nu}{Mc}.$$

(2) The recoil energy of the atomic nucleus is

$$K = \frac{1}{2}Mv^2 = \frac{1}{2}M\left(\frac{h\nu}{Mc}\right)^2 = \frac{(h\nu)^2}{2Mc^2}.$$

(3) Evaluating the ratio of K to $E = h\nu_0$, we obtain

$$\frac{K}{E} = \frac{(h\nu)^2/2Mc^2}{h\nu_0} \approx \frac{h\nu}{2Mc^2} = \frac{1}{2} \times \frac{1}{4 \times 10^6} = \frac{1}{8 \times 10^6}.$$

(4) From Eq. (6.4), we have

$$\nu = \sqrt{\frac{1 + (v/c)}{1 - (v/c)}}\nu_0 \approx \sqrt{\left(1 + \frac{v}{c}\right)^2}\nu_0 = \left(1 + \frac{v}{c}\right)\nu_0.$$

Thus, we find the expression for the ratio $\Delta \nu / \nu_0$ as follows:

$$\frac{\Delta \nu}{\nu_0} = \frac{\nu - \nu_0}{\nu_0} = \frac{(v/c)\nu_0}{\nu_0} = \frac{v}{c} = \frac{gH}{c^2}.$$

∎

Advanced Problems

Problem 6.2.   Theory of special relativity and its application to GPS

According to the theory of special relativity, a couple of clocks moving relative to each other, as one in a moving train and another at rest on the ground, follow different time progress. In the following, we will derive fundamental equations in this theory and apply them to the problem of positioning by a car navigator. For this purpose, we will consider experiments in which sound and light travel on a vehicle moving relative to the ground. In such a description, we denote the time and space measured by the clock and ruler placed on the ground as (t, x) and those measured by apparatus carried by the moving vehicle as (t', x').

I In Secs. I and II, we will consider a sound wave, which propagates through air at speed V. We assume in these two sections that observers of experiments instantaneously detect flashes of lamps because the light propagates much faster than the sound.

Suppose there is a vehicle with two carts combined by a solid long bar as shown in Fig. 6.3. It moves rightward along the x-axis at uniform velocity v. Each cart carries a set of a loudspeaker and a lamp and let the distance between the two loudspeakers be L. When the loudspeaker on the left cart passes by a point O, a person on the left cart sends a sound pulse and simultaneously flashes the left lamp. That instant defines the time $t = 0$ of the clock on the ground. One end of a ruler, which lies on the x-axis on the ground, is placed at the point O, which defines the origin of the coordinate $x = 0$. At the instant when the person standing on the right cart detects the sound pulse from the left

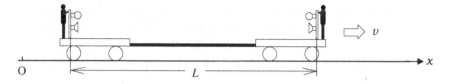

Fig. 6.3.

loudspeaker, he flashes the lamp and simultaneously sends a sound pulse from the right loudspeaker to the left cart. The first person on the left cart flashes the lamp upon detecting the sound pulse from the right cart.

An observer standing on the ground observes this experiment. To solve the problems in Sec. I, assume that the air between the two carts remains at rest on the ground, so that the sound propagates between the carts at speed $V(> v)$ relative to the ground.

(1) When does the observer detect the flash of the right lamp? Let this time be t_1 as measured by the clock on the ground. Express the time t_1 in terms of L, V and v.

(2) When does the observer detect the second flash of the left lamp? Let the time be t_2 as measured by the clock on the ground. Express the time t_2 in terms of L, V and v.

II In the next experiment a large and long box is set on the vehicle, as shown in Fig. 6.4. The two persons in the box manipulate the lamps and loudspeakers in the same manner as those in the previous experiment. The whole system moves rightward at uniform velocity v. Because the air enclosed in the box is at rest relative to the box, the sound pulses propagate at speed V relative to the box. The time $t' = 0$ as measured by the clock in the box is defined by the instant when a pulse of sound is sent from the

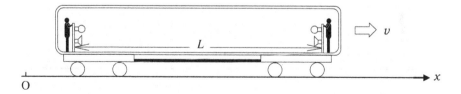

Fig. 6.4.

left loudspeaker. One end of a ruler that defines the origin of coordinate $x' = 0$ is fixed at the position of the left loudspeaker. An observer standing in the box observes this experiment.

(3) When does the observer detect the flash from the right lamp? Let this time be t'_1 as measured by the clock in the box. Express the time t'_1 in terms of L, V and v or a part of these three quantities.

(4) When does the observer detect the second flash from the left lamp? Let the time be t'_2 as measured by the clock in the box. Express the time difference $t'_2 - t'_1$ in terms of L, V and v or a part of these three quantities.

(5) The time $t = 0$ as measured by the clock placed on the ground is defined by the instant when the sound pulse is sent by the left loudspeaker. The location of the left loudspeaker at that moment defines $x = 0$ of the ruler placed on the ground. When does the right lamp flash? Let the time be t_1. Show that t_1 equals t'_1.

III The result of part (5) implies that both clocks, the one in the box and the other on the ground, follow the same time progress. However, Albert Einstein showed that two clocks should follow different time progress, if the two clocks are moving relative to each other.

We assume that a lamp and a mirror are placed at the left and right ends on the vehicle, respectively, as shown in Fig. 6.5 and the vehicle moves rightward at uniform velocity v. The distance between the lamp and the mirror is L as measured by the ruler placed on the ground.

At the instant when the lamp passes the point O, the left person flashes the lamp. This instant defines $t = 0$ and the

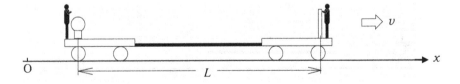

Fig. 6.5.

position of O defines $x = 0$ as measured by the clock and the ruler fixed on the ground, respectively. The pulse propagates rightward and is reflected by the mirror at the right end, coming back to the position of the lamp.

In Sec. III, we consider the case in which this experiment is observed by a clock and a ruler fixed on the ground, while in Sec. IV the case in which observation is made by a clock fixed to the vehicle. An important point of the Einstein's theory is that the speed of light relative to either observer has the same value about 3×10^8 m/s. This universal speed of light is denoted by c, hereafter.

(6) When does the light pulse arrive at the mirror? Let this time be t_1 as measured by the clock on the ground. Express t_1 in terms of L, v and c.

(7) When does the light pulse reflected by the mirror come back to the position of the lamp? Let this time be t_2 as measured by the clock placed on the ground. Express the time t_2 in terms of L, v and c.

IV We now analyze this experiment in terms of the time measured by the clock carried by the vehicle. The moment when the lamp flashes defines the time $t' = 0$ as measured by this clock. Let the time when the light pulse arrives at the mirror be t_1' and the time when the pulse comes back to the position of the lamp be t_2'. Because the light pulse goes to the mirror and comes back to the lamp on the vehicle at the same speed c, t_2' should be twice of t_1'. Accordingly, if the two clocks, the one on the cart and the other on the ground, follow the same time progress, the relation $t_2 = 2t_1$ should hold. However, the result of part (7) implies that it is not the case and the two clocks progress in different manners.

Then, Albert Einstein thought that the time indicated by the clock on the vehicle should not be equal to that by the clock on the ground, when the two clocks relatively move to one another, and he assumed that there is a linear relation between t and t':

$$t' = at + b(x - vt), \qquad (6.5)$$

where a and b are constants. In addition, he considered that the scale of a ruler on the vehicle is possibly different from that on the ground and he assumed that there is a linear relation as follows:

$$x' = at + \beta(x - vt), \qquad (6.6)$$

where the origin of coordinate of x' is defined by the position of the lamp.

Now we determine the constants a, b, α and β from the following arguments about an experiment on the vehicle shown in Fig. 6.5.

Because the lamp is located at $x = vt$ in terms of the time and the coordinate measured by the clock and the ruler fixed on the ground and at $x' = 0$ in terms of the coordinate system moving with the vehicle, we find from Eq. (6.6) that $\alpha = 0$.

(8) Using Eq. (6.5) and the relation $t'_2/t'_1 = 2$, derive an expression for b in terms of a, c and v.

(9) Let a light pulse be emitted at $t = t' = 0$ from the lamp. We first follow this pulse with the clock and ruler fixed to the vehicle. Suppose at time t' the light pulse passes a point denoted by P, which is located at x'. Then, we have the relation

$$x' = ct'. \qquad (6.7)$$

If we follow the same light pulse with the clock and ruler fixed to the ground and denote the position P by x and the time of arrival at P by t, then we have the relation

$$x = ct. \qquad (6.8)$$

Referring to this experiment, derive an expression for β in Eq. (6.6) in terms of a, c and v.

(10) Consider another vehicle that moves at velocity $-v$ relative to the vehicle shown in Fig. 6.5. The time and the coordinate measured by the clock and ruler carried by the new vehicle are denoted by t'' and x'', respectively. Derive expressions for t'' and x'' in terms of t', x', v, a and c. To derive the expressions, use the relations for β and b found in parts (8) and (9).

(11) Because the new vehicle is at rest relative to the ground, we should have the identical relations $t = t''$ and $x = x''$. Using this fact, derive an expression for a in terms of c and v. Then, derive the relations

$$t' = \frac{t - (v/c^2)x}{\sqrt{1 - (v/c)^2}}, \quad x' = \frac{x - vt}{\sqrt{1 - (v/c)^2}}. \qquad (6.9)$$

(12) Let z be a quantity sufficiently small as compared with unity, then we can use the approximation $(1+z)^p \cong 1 + pz$. Assuming that $(v/c)^2$ is sufficiently small as compared with unity, use this approximation for Eq. (6.9) to express t' and x' by the sum of a term without c and a term proportional to c^{-2}.

V The Global Positioning System (GPS) is a system that provides information about our position on the ground using data conveyed by electromagnetic waves from satellites. Car navigators and some cellular phones are its terminals by which we can know our positions. Here, we consider an application of the theory of special relativity to this positioning system.

The car navigator communicates simultaneously with four satellites for positioning in the three dimensional space. Here, however, we consider a simplified system as shown in Fig. 6.6. In this model, the two satellites move rightward with the same velocity v in the same orbit as the car does. Suppose two pulses of the radio wave, the one transmitted by the satellite 1 located at position x_1 at time t_1 and the other transmitted from the satellite 2 located at position x_2 at time t_2, arrived at the car navigator simultaneously. These positions and times, t_1, t_2, x_1 and x_2 are measured by the clock and ruler on the ground. If the car navigator receives these pulses at position x and at time

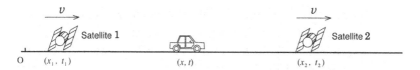

Fig. 6.6.

t, we have

$$|x - x_1| = c(t - t_1), \quad |x - x_2| = c(t - t_2). \tag{6.10}$$

If we solve these two equations for x, we can obtain the position of the car. In the following discussion, restrict yourselves to the case that $x_1 \leq x \leq x_2$ and insert the value $v = 3.8 \times 10^3$ m/s for the velocity of the satellite in numerical calculations.

(13) In the GPS, clocks play a significant role. Clocks are carried not only by satellites but also by car navigators. Therefore, instead of solving the coupled Eqs. (6.10) we may substitute the time t_n as measured by the clock in the car navigator for t in one of the two equations in (6.10) and solve the resultant single equation. How accurate should we measure the time t_n by the clock in the navigator in order to keep the error in a solution for x less than 1 m? Show the upper limit of the allowed error in t_n.

(14) An error in time by the radio clock is typically 1 ms because it uses time signal transmitted from an antenna several hundred kilometers away. It is much larger than the upper limit of error in time required in part (13). Actually, the GPS uses atomic clocks installed in satellites, which achieve an accuracy around 1 ns (1 ns = 10^{-9} s). Solve the coupled equations (6.10) for the position x of the car in terms of x_1, t_1, x_2, t_2 and c. Then, estimate an uncertainty in the computed value of x.

(15) A pulse signal from the satellite 1 is tagged with time t_1' when the pulse is transmitted and with position x_1' of the satellite at that moment. Values of quantities x_1 and t_1 in Eq. (6.10) are calculated from (t_1', x_1') by the inverse transformation of the formulae in Eq. (6.9). Quantities x_2 and t_2 in Eq. (6.10) are similarly calculated by information on x_2' and t_2' from the satellite 2. Two pulses, the one tagged as $(t_1', 0)$ and the other tagged as (t_2', L), arrive at the car navigator simultaneously. Derive an expression for x in terms of t_1', t_2', L, c and v. The relativistic effect should be considered in the approximation employed in part (12).

(16) Suppose pulses transmitted from the two satellites at $t'_1 = t'_2 = 0$ arrive at the car navigator simultaneously. Then the position of the car navigator is

$$x = \frac{L}{2},$$

if the relativistic correction is not considered. Actually, this does not represent the true position of the car navigator. Evaluate the error of position in meter when $L = 2.4 \times 10^4$ km.

The result of part (16) implies that, if the GPS does not include the relativistic effect, it will generate practically serious errors. Actually, because electromagnetic waves between satellites and car navigator are influenced by gravitational field of the earth and rotation of the earth, the GPS should be planned to inform our position, taking into account the effect of the general relativity as well.

(the 2nd Challenge)

Solution

(1) Because the vehicle moves a distance vt_1 in time t_1, the position of the right loudspeaker at time t_1 is $vt_1 + L$. For this period, the sound propagates the distance Vt_1. Therefore, we have

$$Vt_1 = vt_1 + L.$$

By solving this relation for t_1, we find

$$t_1 = \frac{L/V}{1 - v/V}.$$

(2) During the time $t_2 - t_1$, the vehicle moves rightward the distance $v(t_2 - t_1)$. For this period, the sound pulse propagates the distance $L - v(t_2 - t_1)$. Hence, we have

$$V(t_2 - t_1) = L - v(t_2 - t_1).$$

Then we can obtain

$$t_2 - t_1 = \frac{L/V}{1 + v/V}.$$

We insert the expression for t_1 derived in part (1) into the left-hand side of this equation to find

$$t_2 = \frac{L/V}{1 - v/V} + \frac{L/V}{1 + v/V} = \frac{2L/V}{1 - (v/V)^2}.$$

(3) Because, during the time t'_1, the sound propagates the distance L at speed V, we have

$$t'_1 = \frac{L}{V}.$$

(4) Because it takes the time $t'_2 - t'_1$ for the sound to propagate the distance L at speed V, we find

$$t'_2 - t'_1 = \frac{L}{V}.$$

(5) Because the sound propagates at speed $V + v$ relative to the ground during the time t_1, we replace V in the result of part (1) with $V + v$ to obtain

$$t_1 = \frac{L}{V},$$

which is the same expression as that for t'_1 derived in (3).

(6) By replacing V in the result of part (1) with c, we find that

$$t_1 = \frac{L/c}{1 - v/c}.$$

(7) Replacement of V in the result of part (2) with c yields

$$t_2 = \frac{L/c}{1 - v/c} + \frac{L/c}{1 + v/c} = \frac{2L/c}{1 - (v/c)^2}.$$

(8) By substituting

$$t'_1 = at_1 + b(x_1 - vt_1), \quad t'_2 = at_2 + b(x_2 - vt_2).$$

into $2t'_1 = t'_2$, we can get

$$2[at_1 + b(x_1 - vt_1)] = at_2 + b(x_2 - vt_2). \tag{6.11}$$

The times, t_1 and t_2, are given by parts (6) and (7), respectively. The coordinates, x_1 and x_2, being measured by the ruler fixed on the ground, are those of the mirror at t_1 and of the lamp at t_2, respectively. Therefore, we have $x_1 = L + vt_1$ and $x_2 = vt_2$. Substitution of these relations into Eq. (6.11) yields

$$2(at_1 + bL) = at_2.$$

If we substitute the expressions for t_1 and t_2 obtained in (6) and (7) into this equation, we obtain

$$\frac{vL/c^2}{1 - (v/c)^2} a + Lb = 0.$$

Then, we find

$$b = -\frac{v/c^2}{1 - (v/c)^2} a. \tag{6.12}$$

(9) Because $\alpha = 0$, substitution of Eqs. (6.5) and (6.6) into Eq. (6.7) yields

$$\beta(x - vt) = c[at + b(x - vt)].$$

Then we substitute $x = ct$ into this equation and get

$$\beta = \frac{c}{c - v}[a + b(c - v)] = \frac{c^2}{c^2 - v^2} a, \tag{6.13}$$

where b was eliminated by using Eq. (6.12).

(10) The expressions for b and β in Eqs. (6.12) and (6.13) are substituted into Eqs. (6.5) and (6.6). Then we obtain

$$x' = \frac{a(x - vt)}{1 - (v/c)^2} \quad \text{and} \quad t' = a\frac{t - (v/c^2)x}{1 - (v/c)^2}. \tag{6.14}$$

Replacement of x, t, x', t' and v in the above expressions with x', t', x'', t'' and $-v$ yields

$$x'' = \frac{a(x' + vt')}{1 - (v/c)^2} \quad \text{and} \quad t'' = a\frac{t' + (v/c^2)x'}{1 - (v/c)^2}.$$

(11)
$$t = t'' = a\frac{t' + (v/c^2)x'}{1 - (v/c)^2}$$

$$= a\left[a\left(t - \frac{vx}{c^2}\right) + a\left(\frac{v}{c^2}\right)(x - vt)\right]\frac{1}{[1 - (v/c)^2]^2}$$

$$= \frac{a^2}{1 - (v/c)^2}t.$$

Comparing the first term with the final term, we obtain

$$a = \sqrt{1 - (v/c)^2}. \tag{6.15}$$

If we insert this expression of a into Eq. (6.14), we find

$$t' = \frac{t - (v/c^2)x}{\sqrt{1 - (v/c)^2}} \quad \text{and} \quad x' = \frac{x - vt}{\sqrt{1 - (v/c)^2}}.$$

Note: The relation of Eq. (6.15) can be derived in an alternative way as follows:

$$x = x'' = \frac{a(x' + vt')}{1 - (v/c)^2}$$

$$= \frac{a^2}{\{1 - (v/c)^2\}^2}\left[(x - vt) + v\left(t - \frac{v}{c^2}x\right)\right] = \frac{a^2}{1 - (v/c)^2}x.$$

The comparison of the first and the last term yields Eq. (6.15).

(12) According to the formula presented in the problem, we find

$$\frac{1}{\sqrt{1-(v/c)^2}} \cong 1 + \frac{v^2}{2c^2}.$$

Substitution of this into the right-hand side of Eq. (6.9) yields

$$t' = \frac{t-(v/c^2)x}{\sqrt{1-(v/c)^2}} = \left(t - \frac{v}{c^2}x\right)\left[1 + \frac{1}{2}\left(\frac{v}{c}\right)^2\right]$$

$$= t - \frac{v}{c^2}x + \frac{1}{2}\left(\frac{v}{c}\right)^2 t, \tag{6.16}$$

$$x' = \frac{x-vt}{\sqrt{1-(v/c)^2}} = (x - vt) + \frac{1}{2}\left(\frac{v}{c}\right)^2(x - vt). \tag{6.17}$$

(13) Substitution of $t = t_n$ into the left equation of Eq. (6.10) yields

$$x = x_1 + c(t_n - t_1).$$

Hence, if the error $|\Delta t|$ is generated in the measurement of t_n, an error $|\Delta x|$ of the calculated value of x becomes

$$|\Delta x| = |c\Delta t|, \tag{6.18}$$

which follows from Eq. (6.18) and the requirement $|\Delta x| < 1\,\mathrm{m}$ that

$$\Delta t < \frac{1\,\mathrm{m}}{3.0 \times 10^8\,\mathrm{m/s}} = 3.3 \times 10^{-9}.$$

(14) If we solve the coupled equations

$$x - x_1 = c(t - t_1) \quad \text{and} \quad x_2 - x = c(t - t_2),$$

we obtain

$$x = \frac{x_1 + x_2}{2} - \frac{c}{2}(t_1 - t_2). \tag{6.19}$$

If t_1 and t_2 measured by atomic clocks in satellites involve the uncertainty $|\Delta t| = 1\,\mathrm{ns}$, the uncertainty in the position computed by Eq. (6.19) is estimated as

$$|\Delta x| = c|\Delta t| = 3.00 \times 10^8 \times 1.0 \times 10^{-9} = \underline{3.0 \times 10^{-1} \text{m}}$$

(15) Because a car moves much more slowly than satellites, we may employ the approximation in which the car is at rest on the ground and the satellite moves with velocity v relative to the car. Then, using an inverse transformation of Eq. (6.9) and the approximation in part (12), we obtain

$$x \approx (x' + vt') \left[1 + \frac{1}{2} \left(\frac{v}{c} \right)^2 \right] \quad \text{and}$$

$$t \approx \left(t' + v \frac{x'}{c^2} \right) \left[1 + \frac{1}{2} \left(\frac{v}{c} \right)^2 \right].$$

Substitution of $x' = 0$ and $t' = t'_1$ into these two equations yields

$$x_1 = vt'_1 \left[1 + \frac{1}{2} \left(\frac{v}{c} \right)^2 \right] \quad \text{and} \quad t_1 = \left[1 + \frac{1}{2} \left(\frac{v}{c} \right)^2 \right] t'_1,$$

and substitution of $x' = L$ and $t' = t'_2$ yields

$$x_2 = (L + vt'_2) \left[1 + \frac{1}{2} \left(\frac{v}{c} \right)^2 \right] \quad \text{and}$$

$$t_2 = \left(t'_2 + v \frac{L}{c^2} \right) \left[1 + \frac{1}{2} \left(\frac{v}{c} \right)^2 \right].$$

If we substitute these four expressions into Eq. (6.19), we find that

$$x = \underline{\frac{L}{2} \left[1 + \frac{v}{c} + \frac{1}{2} \left(\frac{v}{c} \right)^2 \right] + \left(\frac{v-c}{2} t'_1 + \frac{v+c}{2} t'_2 \right) \left[1 + \frac{1}{2} \left(\frac{v}{c} \right)^2 \right]}$$

(16) Substitution of $t'_1 = t'_2 = 0$, $v = 3.8 \times 10^3$ m/s and $L = 2.4 \times 10^4$ km into the results in part (15) yields

$$x = \frac{2.4 \times 10^7}{2} \times \left(1 + \frac{3.8 \times 10^3}{3.0 \times 10^8}\right) = \frac{L}{2} + \underline{152\,\text{m}},$$

up to the first order in (v/c).

Hence, the position of the car deviates by 152 m from the position predicted by the non-relativistic theory.

■

Problem 6.3. The Bohr model and super-shell

In the 17th century, Sir Isaac Newton formulated laws on motion of a particle in a wide range of scales, from "the fall of an apple" on the earth to the motion of planets in the solar system. It turned out at the beginning of the 20th century that Newton's laws cannot be applied to motions in the microscopic world, i.e., the objects in atomic scale. In order to describe and to predict such motions, it has been recognized that a totally new theory would be necessary.

In these circumstances, a lot of physicists contributed to establish the quantum theory. In particular, N. Bohr played a leadership role at the early stage. By introducing the concept of the **stationary state** as well as the **frequency condition**, he constructed what is called the **semi-classical quantum theory**, which explained the experimental results on atomic spectra. Today, the quantum theory has been recognized as a theory that describes a lot of properties of materials such as color, conductance, and hardness from a microscopic point of view.

In this problem, we consider the motion of an electron in a microscopic system on the basis of the semi-classical quantum theory. For that purpose, we denote the mass and the charge of an electron by m and $-e$, where e is the elementary electric charge. We first apply Bohr's quantization condition and, if necessary, the generalized quantization condition proposed by Sommerfeld, Wilson, and Ishihara, to a microscopic system.

I First, let us consider a hydrogen atom, using the Bohr model. In that model, the nucleus (proton) is a positive point charge located at the center of the atom, in which an electron, a negative point charge, moves in a circular orbit around the nucleus at a constant speed. In the quantum theory, all objects exhibit properties of waves as well as those of particles. This kind of wave is called the **matter wave**. The wavelength of the matter wave (called the **de Broglie wavelength**) is proportional to the inverse of the momentum of the particle and its proportional coefficient h is known as the **Planck constant**.

The quantization condition given by Bohr is equivalent to the statement that in the circular motion of the electron the circumference should be the de Broglie wavelength λ multiplied by a positive integer. The positive integer denoted by n is called the **quantum number** and the electronic state specified by the quantum number n will be called the **n-th stationary state**, hereafter.

(1) Write down the relation between the de Broglie wavelength λ_n and the radius r_n of the circular orbit of an electron in the n-th stationary state.

(2) Noticing that the electrostatic force between the nucleus and the electron plays the role of the centripetal force, find an expression for r_n. Denote the proportional constant in Coulomb's law by k_0.

(3) Using r_n acquired in the above question, find an expression for the energy of the electron E_n in the n-th stationary state.

Bohr succeeded in calculating the spectral series of the hydrogen atom. He assumed that the hydrogen atom will emit a photon when the electronic state changes from a higher energy state to a lower energy state and the energy of the emitted photon is equal to the difference of energies of the electron before and after the transition. If we denote the energies before and after the transition by $E_{n'}$ and E_n, respectively, the frequency of what is called the Lyman series of the spectrum is derived by substitution of the expression of energy derived in part (3) into the energy differences $E_{n'} - E_n$ between the states with

$n' = 2, 3, 4, \ldots$ and with $n = 1$. In the same manner, the Balmer series of the spectrum can be described by the energy differences between the states with $n' = 3, 4, 5, \ldots$ and with $n = 2$. It is verified by these results that the quantum theory is very useful.

II Next, we consider an electron that moves at a constant speed along the x-axis back and forth between two walls located at $x = 0$ and $x = L\, (L > 0)$. We employ the condition that $n\lambda = 2L$ with the positive integer n. This is analogous to the Bohr quantization condition, because the twice of the distance between the two walls, which is the magnitude of displacement during one lap of the motion, corresponds to the circumference in the case of the hydrogen atom.

(4) Find an expression for a speed v_n of an electron in the n-th stationary state.

(5) Find an expression of the energy E_n of an electron in the n-th state. Note that, because the potential energy is constant in this case, we need not add it into the energy E_n.

Ordinarily, the pressure of a gas in a box is explained in terms of the Newtonian mechanics, whereas the force exerted on the wall by electrons can be dealt with the quantum theory as follows. Assume that the wall at $x = L$ is movable. If this wall moves slowly, we can assume that the electron will stay in the same stationary state. When the wall at $x = L$ moves to $x = L + \Delta L$ (ΔL is a small distance) and the energy of the electron in the n-th stationary state changes from E_n to $E_n + \Delta E_n$, the force F_n with which the electron pushes the wall is given by

$$F_n = -\frac{\Delta E_n}{\Delta L}. \tag{6.20}$$

(6) Using Eq. (6.20), express F_n in terms of v_n, m and L.

Let us consider another way to find stationary states of an electron. Generally, motion of a particle on the x-axis can be specified by $x(t)$ and $p(t)$, which represent the time evolution of the position and the momentum of the particle, respectively. Then, consider the trajectory of the point $(x(t), p(t))$ on the plane in which we take the x-axis in the horizontal direction and the p-axis in the vertical direction. Such a plane is called a **phase space** and the point at $(x(t), p(t))$ is called the **representative point** of the particle. Motion of a particle in the ordinary space can be represented by the motion of its representative point in the phase space. For example, as shown in Fig. 6.7, the instantaneous change of momentum of an electron from $p = -mv$ to $p = mv$ on collision with the wall at $x = 0$ can be represented by a jump of the representative point (a segment in Fig. 6.7) from $(0, -mv)$ to $(0, mv)$ in the phase space.

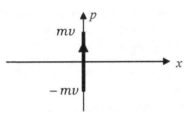

Fig. 6.7.   The trajectory in the phase space which represents the instantaneous collision.

We revisit the motion of the electron that travels back and forth between two walls located at $x = 0$ and L at the constant speed v.

(7) Draw the closed trajectory of the representative point of the electron in the phase space.

(8) Calculate the area of the region enclosed by the trajectory drawn in part (7).

(9) The electron is in the n-th stationary state considered in part (4). Prove that the area S_n enclosed by the trajectory of the representative point in the phase space is equal to the Planck constant multiplied by n.

III Let us generalize the result derived in part (9) as follows: when the trajectory in the phase space forms a closed loop, a quantum stationary state is acquired from the condition that the area enclosed by the trajectory is equal to the Planck constant multiplied by a positive integer. This condition is called the **Bohr-Sommerfeld quantization condition**.

The quantization condition of Bohr stated above implies that a standing wave is formed in hydrogen atom or in the space between the two walls and it can be applied to the motion at a constant speed. On the other hand, the Bohr-Sommerfeld quantization condition can be applied to more generalized situations.

As an example, we consider a harmonic oscillation of an electron on the x-axis. In this motion, since the de Broglie wavelength of the electron as well as the momentum varies with time, it is difficult to consider a standing wave. On the other hand, the Bohr-Sommerfeld quantization condition can be directly applied to this motion because the motion of the electron describes a closed trajectory in the phase space.

(10) The potential energy of this electron at position x is expressed as $\frac{1}{2}kx^2$, where k is a positive constant. Find an expression for the amplitude of the harmonic oscillation of the electron with the mechanical energy of E.

(11) Write down the expressions for position $x(t)$ and the momentum $p(t)$, assuming that $x(t)$ takes the maximum value at $t = 0$.

(12) Draw the trajectory of the representative point of the electron in the phase space.

(13) Apply the Bohr-Sommerfeld quantization condition to the trajectory drawn in part (12) to find an expression for the energy E_n of the electron in the n-th stationary state.

IV On the basis of the results derived in Sec. III, let us examine whether the Thomson model can be applied to calculating the energy levels of the hydrogen atom. According to the Thomson model, suppose an electron moves in a sphere in which positive charges are uniformly distributed. Taking the origin of coordinates at the center of the charged sphere so that the the x-, y- and z-axes are perpendicular to each other, then we can express the potential energy of an electron located at a point (x, y, z) as $\frac{1}{2}K(x^2 + y^2 + z^2)$. Here, K is a positive constant and $\sqrt{x^2 + y^2 + z^2}$ is the distance from the center of the charged sphere to the electron. This expression of the potential energy implies that the motion of an electron in the Thomson model is a superposition of harmonic oscillations along the three axes. We represent each of the harmonic oscillations by a trajectory in each of the phase spaces of (x, p_x), (y, p_y), and (z, p_z), where p_x, p_y and p_z are the x- y- and z-components of the momentum, respectively. Finally, the Bohr-Sommerfeld quantization condition is applied to each of the three trajectories.

(14) Write down an expression for the energy E of an electron in terms of p_x, p_y, p_z, x, y and z.

(15) Apply the Bohr-Sommerfeld quantization condition to find the expression for the quantized energy $E(n_x, n_y, n_z)$ of the electron. Here n_x, n_y and n_z are the quantum numbers of the motion along the x-, y- and z-axes, respectively.

This result is clearly different from expression for energy acquired in part (3), which agrees with the experimental results. Therefore, we find that the Thomson model cannot be applied to calculating the energy level of the hydrogen atom. On the other hand, Nagaoka proposed another model of atom (1904) in which electrons move around a positively charged sphere. The electric field outside the charged sphere is the same as that produced by the charge concentrated on the center of the sphere and equivalent to the electric field in an atom of the Bohr model. Hence, the Nagaoka model leads us to the same result as that in part (3) and this model describes the spectrum series of the hydrogen atom.

In a case of a heavy nucleus, however, a stationary state more consistent with experiments is given by considering the heavy nucleus to be a positive charged sphere with finite size rather than to be a point charge, as deduced from the quantization condition based on the relativistic wave equation (the Dirac equation). This is because electrons are moving at a speed close to the light one in the innermost shell (K shell) of a heavy nucleus. Since this relativistic effect reduces the radius of the K shell, it is necessary to take the size of the nucleus into account.

V Recently, the Thomson model revived in the study of electronic properties of the 'cluster', which is an aggregate of metallic atoms such as sodium (Na). In the theory of metals, we often employ a model in which the discrete distribution of the positive charges on ions is replaced with a continuous medium of positive charges. Correspondingly, in the theory of the metallic clusters, the aggregate of positive ions of a cluster is replaced with a sphere in which positive charges are uniformly distributed. In addition, we consider that electrons released from the outermost shell in each atom, can freely move in the sphere.

The sodium cluster is different from a hydrogen atom considered in Sec. IV in the number of electrons. We assume that many electrons are uniformly distributed in the sphere of positive charges to keep electrical neutrality of the cluster. Suppose we have a cluster of N atoms. Then, the potential energy of each electron is a sum of the two contributions, the one from the uniformly distributed N positive charges and the other from negatively charged $(N-1)$ electrons. However, their contributions almost compensate each other, and it can be considered that the potential energy of the electron is almost constant in a cluster, which is independent of the size of the cluster. This means that one can consider a model system of a sodium cluster to be an electron system confined in a spherical container of the same size as the cluster.

Let us consider here a periodic motion of an electron within a spherical container with radius R. Since the potential energy is constant, the Bohr quantization condition is applicable to the electron motion, like the free motion of an electron between walls considered in Sec. II. To quantize the electron motion, it is necessary to find out periodic motions of the electron. In this case, the periodic motions may be that in regular polygon-like

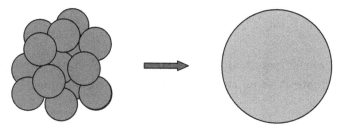

Fig. 6.8. A positively charged sphere model of a metallic cluster, in which positive ions are replaced with positive charges distributed uniformly in a sphere.

orbits. Especially in the sodium cluster, equilateral triangle orbits and square orbits are considered to be important.

(16) Find an expression for the lengths l_3 of one lap of the equilateral triangle and l_4 of one lap of the square in terms of the radius of the sphere R.

(17) Find an expression for the speed u_n of an electron on an equilateral triangle orbit, by using the condition that the length of the orbit is its wavelength multiplied by a positive integer n. By using the same condition, find an expression for the speed w_n on the square orbit.

(the 2nd Challenge)

Solution

The purpose of this problem is to understand the importance of quantum mechanics for describing the states of electrons. Without solving the Schrödinger equation, wave properties of matter, the basic idea of quantum mechanics, can be considered.

(1) The circumference of the circle with radius r_n is $2\pi r_n$. Hence, the Bohr quantization condition is

$$2\pi r_n = n\lambda_n.$$

(2) Assuming that the electron moves at a speed v_n around the nucleus in a circular orbit, the equation of motion is

$$m\frac{v_n^2}{r_n} = k_0\frac{e^2}{r_n^2}.$$

Hence, the momentum of the electron can be calculated as

$$mv_n = \frac{h}{\lambda_n} = \frac{nh}{2\pi r_n},$$

where the result in part (1) has been substituted for λ. Eliminate v_n from two equations above, we obtain

$$r_n = \frac{n^2}{mk_0e^2}\left(\frac{h}{2\pi}\right)^2.$$

(3) The total energy of the electron is the sum of the kinetic energy $\frac{1}{2}mv_n^2$ and the Coulomb potential $-k_0\frac{e^2}{r_n}$. From the equation of motion considered in part (2), the kinetic energy is written as $\frac{1}{2}mv_n^2 = \frac{k_0e^2}{2r_n}$. Hence, we obtain

$$
\begin{aligned}
E_n &= \frac{1}{2}mv_n^2 - k_0\frac{e^2}{r_n} \\
&= -\frac{k_0e^2}{2r_n} \\
&= -\frac{m(k_0e^2)^2}{2}\left(\frac{2\pi}{h}\right)^2\frac{1}{n^2},
\end{aligned}
$$

where the result of part (2) has used to derive the final line.

(4) The quantization condition (which is a condition of standing wave between two walls) is

$$
2L = n\lambda_n.
$$

On the other hand, the momentum of the electron is

$$
mv_n = \frac{h}{\lambda_n}.
$$

Hence, the speed of the electron is

$$
v_n = \frac{h}{m\lambda_n} = \frac{nh}{2mL}.
$$

(5) In this case, the energy involves only the kinetic energy. Hence,

$$
E_n = \frac{mv_n^2}{2} = \frac{n^2h^2}{8mL^2}.
$$

(6) Differentiating $-E_n$ with respect to L, we have

$$
F_n = -\frac{dE_n}{dL} = \frac{n^2h^2}{4mL^3}.
$$

Then, the result of part (4) leads to

$$
F_nL = mv_n^2.
$$

(7) As shown in Fig. 6.9, the trajectory is a rectangular whose apexes are $(0, mv_n)$, (L, mv_n), $(L, -mv_n)$, $(0, -mv_n)$, and the direction is clockwise.

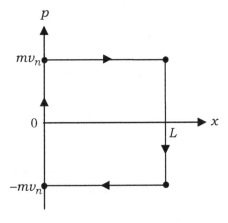

Fig. 6.9.

(8) The area of the region enclosed with the rectangular shown in Fig. 6.9 is

$$L \cdot 2mv_n = \underline{2mv_nL}.$$

(9)
$$S_n = 2mv_n L$$
$$= nh,$$

where we insert the result of part (4) to derive the rightmost hand side.

(10) Since the mechanical energy is

$$E = \frac{1}{2}mv^2 + \frac{1}{2}kx^2.$$

The position x has the maximum value $x_{\max}$ when $v = 0$. Hence, we have

$$x_{\max} = \sqrt{\frac{2E}{k}}.$$

(11) The angular frequency of this harmonic oscillator is $\omega = \sqrt{\frac{k}{m}}$. When $x(t)$ becomes maximum at $t = 0$, $x(t) = \sqrt{\frac{2E}{k}}\cos(\omega t)$. Differentiating $x(t)$ with respect to t, we obtain an expression for $v(t)$. Hence the momentum is

$$p(t) = mv(t) = m\frac{dx(t)}{dt}$$

$$= -m\omega\sqrt{\frac{2E}{k}}\sin(\omega t) = -\sqrt{2mE}\sin(\omega t).$$

(12) As shown in Fig. 6.10, the trajectory is an elliptic orbit where the amplitude of the momentum p is $\sqrt{2mE}$ and that of the position x is $\sqrt{\frac{2E}{k}}$. The direction of the trajectory is clockwise.

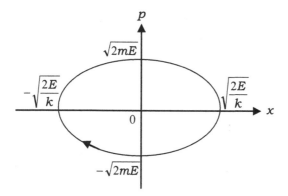

Fig. 6.10.

(13) The area of the ellipse described in part (12) is

$$S = \pi\sqrt{2mE}\cdot\sqrt{\frac{2E}{k}} = 2\pi\sqrt{\frac{m}{k}}E \left(= \frac{2\pi E}{\omega}\right).$$

Hence, the Bohr-Sommerfeld quantization condition can be written as

$$2\pi\sqrt{\frac{m}{k}}E_n = nh.$$

Then the n-th energy value is

$$E_n = n\frac{h}{2\pi}\sqrt{\frac{k}{m}} = n\hbar\omega,$$

where $\hbar = \frac{h}{2\pi}$.

For large n, this is an asymptotic expression for the exact solution of energy values of a harmonic oscillator, which is given by the quantum mechanics:

$$E_n = \left(n + \frac{1}{2}\right)\hbar\omega.$$

(14) Since the motions in three directions are independent of each other, the total energy is given by the sum of their energies:

$$E = \left(\frac{p_x^2}{2m} + \frac{kx^2}{2}\right) + \left(\frac{p_y^2}{2m} + \frac{ky^2}{2}\right) + \left(\frac{p_z^2}{2m} + \frac{kz^2}{2}\right)$$

$$= \frac{1}{2m}(p_x^2 + p_y^2 + p_z^2) + \frac{k}{2}(x^2 + y^2 + z^2).$$

(15) The energy of the x component is $\frac{n_x h\omega}{2\pi}$, where n_x is zero or a positive integer. Similarly, the energies of the y and z components are $\frac{n_y h\omega}{2\pi}$ and $\frac{n_z h\omega}{2\pi}$, respectively. So the total energy is

$$E(n_x, n_y, n_z) = \frac{h\omega}{2\pi}(n_x + n_y + n_z).$$

(16) Since $\overline{AB} = 2 \cdot \overline{OA}\cos(\pi/6) = \sqrt{3}R$ in Fig. 6.11(a), we obtain

$$l_3 = 3\overline{AB} = 3\sqrt{3}R \approx 5.2R.$$

From Fig. 6.11(b), we have

$$l_4 = 4\overline{AB} = 4\sqrt{2}R \approx 5.6R.$$

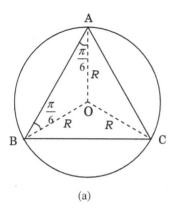
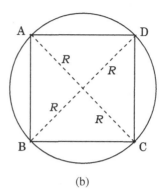

(a)   (b)

Fig. 6.11.

(17) The wavelength of the standing wave made up in the inscribed equilateral triangle is given by $l_3 = n\lambda_{3n}$, and the speed of electrons moving along the inscribed triangle is as follows:

$$u_n = \frac{h}{m\lambda_{3n}} = \frac{nh}{3\sqrt{3}mR}.$$

Similarly, the speed of electrons moving along the inscribed square is as follows:

$$w_n = \frac{nh}{4\sqrt{2}mR}.$$

∎

Problem 6.4.   Fate of the Sun

Discovery of a strange star, white dwarf

The heliocentric theory was widely accepted in the 17th century. If this theory is correct, the earth location in space should significantly change in half a year since the earth moves around the sun in one year. Then, we should expect a slight shift of the fixed-star position accompanied by the earth motion, which is called the **parallax**. However, the parallax was not detected until 19th century when Frederick Bessel performed precise position measurements of nearby stars in order to detect it. Since nearby stars are expected to be bright, he selected many bright stars and detected the parallax for the first time in 1838. Among them, he measured the precise location of Sirius, the brightest star in the constellation of Canis Major, and detected its parallax. He also reported that Sirius showed extra-motion other than the parallax. Based on its extra-motion, they discovered that Sirius is in a binary star system, although they could not see the companion star (1844). Using the state-of-the-art refractor in 1862, Alvan Clark detected a faint star very close to Sirius. He found that two stars showed similar temperatures, but that the bright star (Sirius A) was ten times brighter than the faint star (Sirius B). This fact shows that the surface area of Sirius A is 10^4 times larger than that of Sirius B, which indicates that the radius of Sirius A is 100 times bigger than that of Sirius B. Moreover, the result of the observation of their orbits indicated that the two stars would have similar masses. Based on the astrophysical study, it was known for Sirius A to have a density about $1\,\mathrm{g/cm^3}$ that is similar to that of the sun, whereas Sirius B should have a density about $1 \times 10^6\,\mathrm{g/cm^3}$ that is much denser than any matter known at that time. We will learn that Sirius B is a **white dwarf**. With regard to the density, the gold (Au) is $19.32\,\mathrm{g/cm^3}$ and even the osmium (Os), the densest matter on the earth, is only $22.57\,\mathrm{g/cm^3}$. Arthur Eddington, a famous astronomer in UK, said (1926) 'Apart from the incredibility of the result, there was no particular reason to view the calculation with suspicion'. Let us learn how to explain this incredible result.

Particle motion in a very small scale — Heisenberg uncertainty principle

In Newtonian mechanics, we can specify a state of any particle, like an electron, at some instant by a set of its position and velocity. In the following, we employ particle momentum instead of its velocity. Then, we can specify a state of any particle at a given time by its position r and its momentum p. For simplicity, we consider the situation of one-dimensional configuration along the x-axis, then the particle state can be expressed by a set of position x and momentum p. However, according to the quantum mechanics that can treat the very small scale particle dynamics, we cannot simultaneously determine both x and p precisely. For example, you can only determine the particle state such that the position is between x and $x + \Delta x$ and that the momentum is between p and $p + \Delta p$. Here Δx and Δp are uncertainties of the position and momentum of the particle. The product of Δx and Δp cannot be smaller than a certain physical constant. In other words, if the value of Δx you measure is very small, the value of Δp measured cannot be small. Inversely, if the value of Δp you measure is very small, the value of Δx measured cannot similarly be small either. This is not due to a technological problem but due to the principle of physics, which is called the **Heisenberg uncertainty principle**.

A new type of coordinate, phase space

Here, we introduce the concept of **phase space** that describes the particle state. First of all, we consider the one-dimensional case, where the phase space is described by a plane with the horizontal and vertical axes corresponding to the position x and the momentum p, respectively. Since in Newtonian mechanics we can simultaneously determine the particle position and its momentum at a given time, the particle state at this moment corresponds to a point in the phase space. However, in quantum mechanics, due to the Heisenberg uncertainty principle, any particle state does not correspond to a point but to some volume (area) in the phase space that is called a phase volume. Let us consider the case of electron, proton and neutron. These particles are called **fermions**. If the uncertainties of position and momentum are Δx and Δp, this area, $\Delta x \Delta p$, is equal to $\frac{h}{2}$ where h is the Planck constant. According to the quantum mechanics, the phase volumes of the same kind of fermions cannot overlap one another. In other words, when there are N particles of the same kind of fermion (for example, electron), each particle occupies a phase volume $\frac{h}{2}$, thereby N particles as a whole occupy an area of $N\frac{h}{2}$ in the phase space. Therefore, if there are a large number of electrons (or neutrons or protons), they will occupy a large volume of the phase space. Conversely, a phase volume, ΔV, can accept them up to the number $\frac{\Delta V}{h/2}$.

Degenerate state of electrons

When we employ momentum p instead of velocity v, we should note that the kinetic energy of the particle whose mass is m is expressed by $\frac{p^2}{2m}$. Consider a case that there are N electrons in the position range $-R \leq x \leq R$ in the one-dimensional space. We assume that N is as large as Avogadro's number. When the electron temperature is high, electrons are moving very fast, whence their momenta become large. The electrons sparsely scatter in the phase space as shown in Fig. 6.12(a). However, when the electrons are cooled down, their momenta reduce. We should note that a phase volume which can be occupied by a single electron is constant, independent of the temperature. Furthermore, phase volumes of electrons do not overlap one another in the phase space. Therefore, to make the total momentum (whence the kinetic energy) small at the zero temperature, the electrons occupy a certain range

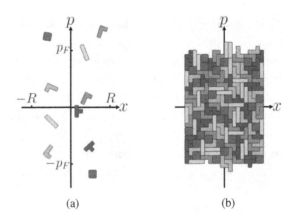

(a)   (b)

Fig. 6.12. Schematic views of the phase space in one-dimension. In this case, the phase space is described in a plane where horizontal and vertical axes correspond to the position, x, and the momentum, p. The phase volume occupied by a single particle has a fixed area while its shape is indefinite. In this figure, there are 5 kinds of shape for simplicity. (a) A non-degenerate state at high temperature. (b) A degenerate state at low temperature.

of the phase space, say from $-p_F$ to p_F, or equivalently $0 \leq |p| \leq p_F$, where p_F is called the **Fermi momentum**.

When the electrons nearly fill the phase space area of $-p_F \leq p \leq p_F$, we refer to this state as the **degenerate state** of electrons. This state is shown in Fig. 6.12(b). When the electrons are in a degenerate state, we will find that the sum of the kinetic energy of all the electrons, U_d, can be expressed in terms of R and N. Answer the following questions.

(1) When N electrons fill the phase space so as to make the total momentum of electrons small, they are in a degenerate state. The electrons will fill the range of the phase space given by $-R \leq x \leq R$ and $-p_F \leq p \leq p_F$. Show that the maximum momentum, p_F, is given by

$$p_F = \frac{hN}{8R}. \qquad (6.21)$$

(2) The kinetic energy of an electron does not depend on its position x, but on its momentum p. Let us consider electrons in the momentum range between p and $p + \Delta p$ and in the position range $-R \leq x \leq R$ where Δp is sufficiently small, such that these electrons can be regarded as having an equal value of kinetic energy, $\frac{p^2}{2m_e}$, where m_e is the electron mass. Calculate the sum of the kinetic energy of the electrons in this range, ΔU_d.

(3) The sum of the kinetic energy of all the electrons can be expressed as $U_d = \sum \Delta U_d$. This sum can be performed by an integration over the momentum range $-p_F \leq p \leq p_F$. Express U_d in terms of R and N by taking into account Eq. (6.21).

Degenerate pressure of electrons

Due to the Heisenberg uncertainty principle, the entire kinetic energy of the electrons, U_d, cannot be zero even if the temperature is very low. In this case, if we push the system by an external force to reduce R, we will find U_d to be increased. The increase of U_d is the work done by the external force, which means that there is a pressure,

$P = -\frac{dU_d}{dR}$. This pressure is called the **degenerate pressure**. If there are protons which are also degenerate, then they can also produce a degenerate pressure. It should be noted that different kinds of particles like a group of electrons and that of protons can overlap with each other in the phase space. Therefore, the degenerate pressure is determined for each kind of particle. It is separately calculated for each species of particle.

(4) There are an equal number of electrons and protons in a volume. Both electrons and protons are in degenerate states. The degenerate pressure of electrons is higher than that of protons. How high is it?

Relativistic and non-relativistic kinetic energy

When the speed of a particle approaches that of light in vacuum, c, we have to include the relativistic effect. When the speed of the particle is much smaller than c, we say it is in a **non-relativistic** situation. According to Einstein's relativity theory, the particle energy, E, is expressed as $E = \sqrt{(mc^2)^2 + (pc)^2}$ where m and p are the particle mass and momentum, respectively. The kinetic energy is defined as the difference between the particle energy with momentum p and that with $p = 0$ as follows:

$$\text{kinetic energy} \equiv E - mc^2 = \sqrt{(mc^2)^2 + (pc)^2} - mc^2.$$

We should note that the speed of particle cannot exceed that of light in vacuum, c. When the speed of the particle is much smaller than c, i.e., $pc \ll mc^2$, the situation is non-relativistic. In the other extreme limit, i.e., $pc \gg mc^2$, we say the situation is **ultra-relativistic**.

(5) Show that the kinetic energy of a particle is approximated by $\frac{p^2}{2m}$ in the non-relativistic case and by pc in the ultra-relativistic case. You can use the approximation formula $\sqrt{1 + x} = 1 + \frac{1}{2}x$ for $|x| \ll 1$. In the non-relativistic case, we know $p = mv$, where v is the speed of the particle. Then, the kinetic energy can be expressed as $\frac{1}{2}mv^2$.

Degenerate pressure in the three-dimensional space

Based on the case of the one-dimensional space, i.e., parts (1) through (3), let us consider the case of the three-dimensional space. Since position $r = (x, y, z)$ and momentum $p = (p_x, p_y, p_z)$ are three-dimensional vectors, the corresponding phase space becomes six-dimensional. A single electron will occupy a small volume of position, $\Delta x \Delta y \Delta z$, and a small volume of momentum, $\Delta p_x \Delta p_y \Delta p_z$. Therefore, a single electron will occupy $\Delta x \Delta y \Delta z \Delta p_x \Delta p_y \Delta p_z$, which is equal to a phase space volume of $\frac{h^3}{2}$, in the three-dimensional phase space. When N electrons are packed in a three dimensional sphere of radius R in the degenerate state, we can calculate the total kinetic energy of the electrons, U_d, in terms of N and R, as shown below.

In the three-dimensional space, we know that the ratio of the sphere volume V to the cube of the radius R^3 is $\frac{4\pi}{3}$. Therefore, we will introduce a constant k that simplifies the calculation.

$$k^3 = 2 \times \left(\frac{4\pi}{3} \right)^2 : \text{definition of } k,$$

$$k \approx 3.274: \text{approximate value.}$$

(6) Let us fill the phase space with N electrons by making the total electron momentum as small as possible so that they are in a degenerate state. When the electrons are in the complete degenerate state, they fill the sphere of radius R in the position space and the sphere of radius p_F in the momentum space. Then, find the maximum momentum of electrons, p_F, as a function of N and R.

(7) Electron's kinetic energy does not depend on its position r but on the magnitude of its momentum p. Let us consider the electrons in the phase space, such that the position r is in the range $r \leq R$ and the momentum p is between p and $p + \Delta p$. We assume that Δp is sufficiently small, such that each electron in this phase space can be regarded as having an equal magnitude of the kinetic energy. We should note that the phase

space volume is

$$\frac{4\pi}{3}R^3 4\pi p^2 \Delta p = \frac{3}{2}k^3 R^3 p^2 \Delta p.$$

You can calculate the sum of the kinetic energy, ΔU_d, for all the electrons within this phase volume by referring to part (5). Calculate ΔU_d in both of the non-relativistic and ultra-relativistic cases.

(8) Then, you can calculate the total kinetic energy, U_d, by summing up the kinetic energy over the various values of p. You can carry out the summation over the momentum by an integration over the momentum from 0 to p_F. Employing the results of parts (6) and (7), you obtain U_d as a function of N and R in the form as shown below. Calculate the powers a and b of N and R, respectively, in both of the non-relativistic and ultra-relativistic cases.

$$\text{non-relativistic case: } U_d = \frac{3}{10k^2}\frac{h^2}{m_e}N^a R^b,$$

$$\text{ultra-relativistic case: } U_d = \frac{3}{4k}hcN^a R^b.$$

Fate of the sun

The sun is a typical star in the universe. It is a sphere with radius 7×10^8 m. The central temperature reaches 1.4×10^7 K. In the center of the sun hydrogen fusion continuously occurs. Solar gravity pulls matter towards the center to crash, against the pressure due to the high temperature pushing matter away. The gravity and the pressure have been in well balance over the last 5 billion years. The hydrogen fusion is a process to convert 4 hydrogen nuclei (protons) to 1 helium nucleus. A huge amount of energy is generated through this process to keep the solar center at high temperature. In this case, the hydrogen in center of the sun will totally be converted to helium in 10 billion years after its birth. After that we cannot expect any significant energy generation in its center. It cools down, which implies reduction of the pressure. Eventually the gravity overcomes the pressure, and the sun contracts. Does it contract forever? If something could stop the contraction, what is it?

Gravitational energy of a star

Let us consider the process that a star with uniform density formed by integrating a large number of small pieces of matter. We consider a stellar matter of radius r and mass $m(r)$, then how much is the potential energy of a small mass Δm when it is located at a distance $x(x \geq r)$ away from the center of the star? This energy is called the **gravitational energy**. The gravity F acting on Δm is expressed as

$$F = \frac{Gm(r)\Delta m}{x^2},$$

where G is the gravitational constant. We assume that the gravitational energy is zero at infinity $(x = \infty)$. We move Δm from infinity to the stellar surface $(x = r)$, then the gravitational energy, ΔU_g, which is the work done by an external force against the gravity F, can be calculated by integrating F over x as follows:

$$\Delta U_g = \int_\infty^r F dx = -\frac{Gm(r)\Delta m}{r}.$$

This is the gravitational energy of Δm at the distance r from the center of the star. When the mass density is uniform, we can calculate the gravitational energy of a star with mass $M = m(R)$ and radius R, as shown below.

(9) Suppose Δr is a very small value. The gravitational energy of mass, Δm, is given by the above equation, where Δm is the mass in the volume of the star between spherical surfaces of radii r and $r + \Delta r$. Noting that this volume is given by $4\pi r^2 \Delta r$, calculate the gravitational energy of this volume, ΔU_g, and express the result in terms of G, M, R, r and Δr.

(10) You can calculate the total gravitational energy, U_g, by summing up ΔU_g over r from 0 to R. Find U_g by using the integration and express the result in terms of G, M and R.

Evolution of stars

Let us consider a star (radius R, mass M) whose hydrogen has been completely converted to the helium through nuclear fusion. The helium atom consists of the neutron, the proton and the electron with the number ratio of $1 : 1 : 1$, respectively. Since the neutron and proton have similar masses, they are called the nucleon. The nucleon is much heavier than the electron, which indicates that the atomic mass is almost determined by the number of nucleons. On the contrary, the degenerate pressure is almost determined by the number of electrons, since the electronic degenerate pressure is much higher than that of the nucleon, as seen in part (4). If we denote the ratio of the nucleon number to the electron number by y, we find $y = 2$ for the helium. If the total number of electrons in the star is N, the star mass, M, can be expressed as $M = y m_H N$ where m_H is the nucleon mass. Using this relation, we can rewrite the expression for U_d (derived in part (8)) with M instead of N.

When the star contracts due to gravity, the radius R also reduces. The total energy, U, of the star is the sum of the gravitational energy U_g derived in part (10) and the kinetic energy U_d derived in part (8). We should note that U_g is negative and U_d is positive. If the total energy, $U = U_g + U_d$, has a minimum value at a certain radius R, the contracting star will reach this minimum U value, which is in a stable state. The star of this state is called a **white dwarf**.

(11) Find the radius and the density of a white dwarf with assumptions that the white dwarf has a uniform mass density and the electrons are in the non-relativistic region. In particular, compute numerically the radius and the density for the case when the sun becomes such a white dwarf.

(12) When a star more massive than the sun contracts to form a white dwarf, its radius becomes smaller and its density becomes higher than that of the sun. Consequently, the momentum of the degenerate electrons becomes large. When all the electrons are in the ultra-relativistic region, show that the total energy U is proportional to R^{-1}. Investigate the stability of this star,

and then explain why a star which is heavier than some critical value does not form a white dwarf but collapses.

(13) When the density of a star becomes high, the kinetic energy of the electrons is in the ultra-relativistic case. In this case, as we have considered in part (12), a star whose mass is exceeding some critical value will collapse since the degenerate pressure of the electrons can no longer sustain the gravity. This critical value is called the **Chandrasekhar mass**, named after its discoverer (Subrahmanyan Chandrasekhar). Calculate the Chandrasekhar mass. How many times is it as compared with the solar mass?

Chandrasekhar mass

In the actual white dwarf, the value of y is larger than 2 and the density is not uniform. Furthermore, since not all the electrons are in the ultra-relativistic region, we should note that the precise value of the Chandrasekhar mass is slightly smaller than that calculated here. When the electrons in a star less massive than the Chandrasekhar mass are in the ultra-relativistic region due to their high density, the star will expand, reducing the density. Then, the electrons turn into the non-relativistic region. The star will be stable as a white dwarf calculated in part (11). On the contrary, in a star more massive than the Chandrasekhar mass, the electronic degenerate pressure can no longer sustain the gravity when the star ceases generating energy through fusion reaction. Consequently, the star collapses further and its density increases. Then, a proton will capture an electron to form a neutron. In this way, all the protons turn to neutrons. This is a birth of a **neutron star**.

(14) There is a neutron star whose mass is just larger than the Chandrasekhar mass and all the neutrons are in the non-relativistic region. Referring to the argument of the formation of a white dwarf, estimate the radius of the neutron star. A star whose mass is just smaller than the Chandrasekhar mass will be a white dwarf. How many times is the radius of the white dwarf as compared with that of the neutron star?

Black hole

When a star like the sun terminates its fusion reaction, there is no heat production inside, resulting in the contraction of the star by the gravity. If the star is as massive as the sun, the electronic degenerate pressure will balance with the gravity. It will become a white dwarf. If the star is more massive than the Chandrasekhar mass, the electronic degenerate pressure cannot sustain the gravity. Then, an electron combines with a proton, forming a neutron. The star contains only the neutrons. Since the neutrons also produce a degenerate pressure, its pressure can support the gravity. It becomes a neutron star. Similarly, if the star is more massive than some critical mass, the neutron degenerate pressure can no longer sustain the gravity. If this happens, the star collapses and swallows up everything, resulting in a state called the **black hole**.

Gravitational constant: $G = 6.67 \times 10^{-11} \mathrm{Nm^2/kg^2}$

Planck constant: $h = 6.63 \times 10^{-34} \mathrm{Js}$

light speed: $c = 3.00 \times 10^8 \mathrm{m/s}$

electron mass: $m_e = 9.11 \times 10^{-31} \mathrm{kg}$

nucleon mass: $m_H = 1.67 \times 10^{-27} \mathrm{kg} = 1830 m_e$

Solar mass: $M_{\mathrm{sun}} = 1.99 \times 10^{30} \mathrm{kg}$

Solar radius: $R_{\mathrm{sun}} = 6.96 \times 10^8 \mathrm{m}$

(the 2nd Challenge)

Solution

(1) The N electrons occupy the area of $Nh/2$ in the phase space.

$$2p_F 2R = \frac{h}{2} N \Rightarrow p_F = \frac{h}{2} \frac{N}{4R} = \frac{hN}{8R}.$$

(2) The sum of the kinetic energy is given by

$$\Delta U_d = \frac{\Delta p 2R}{h/2} \frac{p^2}{2m_e} = \frac{2Rp^2}{hm_e} \Delta p.$$

(3) The entire kinetic energy is given by using Eq. (6.21) as follows:

$$U_d = \sum \Delta U_d = \sum \frac{2Rp^2}{hm_e} \Delta p = \frac{2R}{hm_e} \int_{-p_F}^{p_F} p^2 dp = \frac{2R}{hm_e} \frac{2p_F^3}{3}$$

$$= \frac{4R}{3hm_e} \left(\frac{hN}{8R} \right)^3 = \underline{\frac{h^2 N^3}{384 m_e R^2}}.$$

(4) The degenerate pressure P of particles is given by

$$P = -\frac{dU_d}{dR} = \frac{h^2 N^3}{192 m R^3},$$

where m and N are the particle mass and number of particle. If the numbers of particles with different masses are the same, the degenerate pressure of less massive particles is higher than that of more massive particles. Therefore, we find that the degenerate pressure that supports the matter against the gravity is that of electrons. The degenerate pressure of electrons is <u>1830</u> times larger than that of protons.

(5) In the non-relativistic case,

$$\sqrt{m^2 c^4 + p^2 c^2} - mc^2 = mc^2 \left(\sqrt{1 + \frac{p^2}{m^2 c^2}} - 1 \right)$$

$$\approx mc^2 \left(1 + \frac{p^2}{2m^2 c^2} - 1 \right) = \frac{p^2}{2m}.$$

In the ultra-relativistic case,

$$\sqrt{m^2 c^4 + p^2 c^2} - mc^2 = \frac{p^2 c^2}{\sqrt{m^2 c^4 + p^2 c^2} + mc^2} \approx pc.$$

(6) The maximum momentum of electrons is obtained as follows:

$$\frac{4\pi p_F^3}{3} \frac{4\pi R^3}{3} = \frac{h^3}{2} N \Rightarrow k^3 p_F^3 R^3 = h^3 N,$$

$$\Rightarrow p_F = \underline{\frac{h}{kR} N^{1/3}} = \frac{h}{2R} \left(\frac{9N}{4\pi^2} \right)^{1/3}.$$

(7) In the non-relativistic case, we obtain the result in the way similar to part (2).

$$\Delta U_d = \frac{4\pi p^2 \Delta p \frac{4\pi}{3} R^3}{h^3/2} \frac{p^2}{2m_e} = \frac{\frac{3}{2} k^3 R^3 p^2 \Delta p}{h^3/2} \frac{p^2}{2m_e}$$

$$= \frac{3k^3}{2h^3 m_e} R^3 p^4 \Delta p = \frac{16\pi^2}{3h^3 m_e} R^3 p^4 \Delta p.$$

In the ultra-relativistic case, we obtain in the similar way,

$$\Delta U_d \approx \frac{4\pi p^2 \Delta p \frac{4\pi}{3} R^3}{h^3/2} pc = \frac{\frac{3}{2} k^3 R^3 p^2 \Delta p}{h^3/2} pc$$

$$= \frac{3k^3 c}{h^3} R^3 p^3 \Delta p = \frac{32\pi^2 c}{3h^3} R^3 p^3 \Delta p.$$

(8) In the non-relativistic case, the total kinetic energy U_d is given by

$$U_d = \sum \Delta U_d = \sum \frac{3k^3}{2h^3 m_e} R^3 p^4 \Delta p$$

$$= \frac{3k^3}{2h^3 m_e} R^3 \int_0^{p_F} p^4 dp = \frac{3k^3}{10h^3 m_e} R^3 p_F^5$$

$$= \frac{3k^3}{10h^3 m_e} R^3 \left(\frac{h}{kR} N^{1/3} \right)^5 = \frac{3}{10k^2} \frac{h^2}{m_e} N^{5/3} R^{-2}. \quad (6.22)$$

$$\therefore a = \frac{5}{3}, \quad b = -2.$$

In the ultra-relativistic case, U_d is given by

$$U_d = \sum \Delta U_d = \sum \frac{3k^3 c}{h^3} R^3 p^3 \Delta p$$

$$= \frac{3k^3 c}{h^3} R^3 \int_0^{p_F} p^3 dp = \frac{3k^3 c}{4h^3} R^3 p_F^4$$

$$= \frac{3k^3c}{4h^3} R^3 \left(\frac{h}{kR}N^{1/3}\right)^4 = \frac{3}{4k}hcN^{4/3}R^{-1}. \quad (6.23)$$

$$\therefore \ a = \frac{4}{3}, \quad b = \underline{-1}.$$

(9) Noting that the density ρ inside the star is written as $\rho = \frac{3M}{4\pi R^3}$, we can calculate the gravitational energy ΔU_g as follows:

$$\Delta U_g = -\frac{Gm\Delta m}{r} = -\frac{G\frac{4\pi r^3\rho}{3}4\pi r^2\Delta r\rho}{r} = -\frac{16\pi^2 G}{3}\rho^2 r^4\Delta r$$

$$= -\frac{16\pi^2 G}{3}\left(\frac{3M}{4\pi R^3}\right)^2 r^4\Delta r = \underline{-\frac{3GM^2}{R^6}r^4\Delta r}.$$

(10) The entire gravitational energy of the star, U_g, is given as follows:

$$U_g = \sum \Delta U_g = -\frac{3GM^2}{R^6}\sum r^4\Delta r$$

$$= -\frac{3GM^2}{R^6}\int_0^R r^4 dr = \underline{-\frac{3}{5}\frac{GM^2}{R}}. \quad (6.24)$$

(11) Let us calculate the minimum of the entire energy of a star. Using Eqs. (6.22) and (6.24), we can express U as a function of R. Next, we differentiate U with respect to R and set its derivative equal to zero. Then, we have

$$\frac{dU}{dR} = \frac{d}{dR}(U_g + U_d)$$

$$= \frac{d}{dR}\left(-\frac{3GM^2}{5R} + \frac{3}{10k^2}\frac{h^2}{m_e}\left(\frac{M}{ym_H}\right)^{5/3}\frac{1}{R^2}\right) = 0.$$

$$\therefore \ \frac{3GM^2}{5}R = \frac{3}{5k^2}\frac{h^2}{m_e}\left(\frac{M}{ym_H}\right)^{5/3}.$$

From this equation, we obtain R and ρ at the minimum energy of the star as follows:

$$R = \frac{h^2}{k^2 G m_e (y m_{\mathrm{H}})^{5/3}} M^{-1/3}, \tag{6.25}$$

$$\rho = \frac{3M}{4\pi R^3} = \frac{3}{4\pi} \frac{k^6 G^3 m_e^3 (y m_{\mathrm{H}})^5}{h^6} M^2.$$

Let us consider the case of the sun whose mass is given by M_{sun}. The radius will be $R \doteqdot 7.189 \times 10^6 \,\mathrm{m}$ and the density will be $\rho = 1.278 \times 10^9 \,\mathrm{kg/m^3}$.

When a star of the solar mass collapses to a white dwarf and the degenerate pressure of the electrons sustains the gravity, its density is so high that the mass of a sugar cube size will become as large as one ton. Its radius will be similar to that of the earth. The larger the mass is, the smaller the size becomes.

(12) Using Eqs. (6.23) and (6.24), we can express the entire energy U as follows:

$$U = U_g + U_d = -\frac{3GM^2}{5} R^{-1} + \frac{3}{4k} hc \left(\frac{M}{y m_{\mathrm{H}}} \right)^{4/3} R^{-1}$$

$$= \left\{ -\frac{3GM^2}{5} + \frac{3}{4k} hc \left(\frac{M}{y m_{\mathrm{H}}} \right)^{4/3} \right\} R^{-1}. \tag{6.26}$$

We should note that U is proportional to R^{-1}. Therefore, if the coefficient of R^{-1} is positive, U decreases as R increases. Then the star expands. As a result, the density reduces, and the electrons in the star turns from the relativistic region to the non-relativistic region. And so, the star forms a white dwarf which is stable. If the coefficient of R^{-1} is negative, U decreases as R decreases. The star collapses since the gravity can no longer be sustained.

(13)
$$\frac{3GM_{\text{ch}}^2}{5} = \frac{3}{4k}hc\left(\frac{M_{\text{ch}}}{ym_{\text{H}}}\right)^{4/3},$$

$$M_{\text{ch}}^{2/3} = \frac{5hc}{4k(ym_{\text{H}})^{4/3}G},$$

$$M_{\text{ch}} = \left(\frac{5}{4k}\right)^{3/2}\frac{1}{(ym_{\text{H}})^2}\left(\frac{hc}{G}\right)^{3/2}$$

$$= \frac{6.922}{y^2}M_{\text{sun}} \approx \underline{1.73\,M_{\text{sun}}}.$$

In the actual cases, there are some conditions that are different from this simple assumption. For instance, not all the electrons are in the ultra-relativistic region, and there are materials other than helium etc. With these conditions included, the coefficient of 6.92 turns out to be 5.86, resulting in the Chandrasekhar mass to be $1.4\,M_{\text{sun}}$.

(14) In part (11), we obtained the size of a star in which the degenerate pressure of the electrons sustains the gravity. Similarly, we can calculate the size of a star, in which the degenerate pressure of the neutrons sustains the gravity, by substituting $y \to 1$, $m_e \to m_{\text{H}}$, $M \to M_{\text{ch}}$ in Eq. (6.25).

$$R = \frac{h^2}{k^2Gm_e(ym_{\text{H}})^{5/3}}M^{-1/3}$$

$$\Rightarrow \frac{h^2}{k^2Gm_{\text{H}}^{8/3}}M_{\text{ch}}^{-1/3} = \underline{1.037\times10^4\,\text{m}}.$$

Here we employed $M_{\text{ch}} = 1.73\,M_{\text{sun}}$. This value coincides well with the value when we fill the sphere of the star with neutrons.

In the actual condition, relativistic effects play an important role, which we do not include here.

We can calculate the ratio of the radius of a white dwarf whose mass is just below M_{ch} to that of a neutron star whose mass is just above M_{ch}. Setting their mass equal to each other, we have

$$\frac{\text{Radius of white dwarf}}{\text{Radius of neutron star}} = \frac{\frac{h^2}{k^2 G m_e (y m_H)^{5/3}} M^{-1/3}}{\frac{h^2}{k^2 G m_H^{8/3}} M^{-1/3}}$$

$$= \frac{m_H/m_e}{y^{5/3}}$$

$$= 577.4,$$

where we set $y = 2$. The radius of the white dwarf is $\underline{577}$ times larger than that of the neutron star. ∎

PART II

Experiment

Chapter 7

How to Measure and Analyze Data

7.1. Some Hints for Experiments

In this section, we list some useful hints for carrying out measurements quickly and effectively in experimental competitions within time limit, such as in International Physics Olympiad (IPhO) or domestic competitions. Of course, these hints are also useful for experiments in classes at high schools as well as in labs at universities.

(1) Imagine the whole procedure of measurements before making the measurements.

In some of experimental competitions, such as in IPhO, the measurement parameters are not always given in the text. For example, you may not know in advance how quickly the phenomenon in question will occur or how quickly you should measure it. You may not know in advance how small a voltage step you should use in measurements with an electric circuit. You may not know how many millimeters you should shift a mirror in measurements with an optical interferometer. In such experimental competitions, you are required to determine those parameters by yourself prior to making the measurements. Some students will try to do the measurements as precisely as possible by setting the parameters increments (time, voltage, or distance, etc.) very small from the beginning, which will, however, result in waste of time in many cases. First, you should quickly change the parameter across the whole range required, and observe what happens roughly and qualitatively. Then, by considering the time you have in the competition, you should determine how finely you can measure the phenomenon in more detail. In many cases, the phenomenon in question occurs within a limited range of the parameter. If you find such a range with the quick first observation, then, you can concentrate on measuring more precisely only in that smaller range. Outside of such important ranges, you can do the measurement in a rougher manner. This kind of clever approach saves a lot of time, and can help you avoid inconsistent measurements. Furthermore, if, for example, the situation under measurement is symmetric with respect to the origin along the x-axis, you can do the measurement precisely on just the positive x side. It will then be enough to make a rough measurement on the other to confirm the symmetry. This is a very smart approach. If you find (theoretically) that the phenomenon under investigation in an electric circuit occurs in the same way with both positive and negative voltages, you can omit the measurement at one side of the voltage range.

(2) You do not need to conduct each measurement very precisely

For example, to measure the length of an object with a ruler, some students will try to use a precision of $1/10$ of the minimum division of the scale. But it will waste time. You do not need to be so careful, because the general rule in IPhO is that the error in each measurement should be $1/2$ of the minimum division of the scale. Therefore, for a measurement with a ruler having a minimum division of $1\,\mathrm{mm}$, the measurement result may be $21.3 \pm 0.5\,\mathrm{mm}$, in which $0.3\,\mathrm{mm}$ is within the allowable error tolerance and is not so important. The ruler will not always be made with a precision of $1/10$ of the minimum division, so you do not need to spend much time determining the value to the first decimal place in this case.

On the other hand, it is also a general rule in IPhO that the measurement error with a digital instrument is one step of the last digit. For example, if measuring a voltage with a digital multi-meter, when the value on the meter shows $21.3\,\mathrm{V}$, the measurement result is $21.3 \pm 0.1\,\mathrm{V}$. Be sure that the measurement error is $1/2$ of the minimum division in the case of an analog scale (such as a ruler or a stem thermometer), while it is one unit of the last digit of a digital-scale measurement (such as a digital platform scale or a digital multi-meter).

The value shown on the instrument is sometimes unstable; that is, the last digit is always changing. The value may be around 21.3 V, but the first decimal digit fluctuates between 1 and 5. In this case, the range of the fluctuation should be indicated as the measurement error. So the measurement result in this case is 21.3 ± 0.2 V. Similarly, when the indicator in an analog meter fluctuates in a specific range, you can put the fluctuation extremes as the measurement error. Of course, you should first try your best to suppress the instability in the measurements. But if the fluctuation cannot be eliminated in spite of such efforts, you can regard it as noise and mark it as measurement error. If the amount of fluctuation is acceptable, you should not concentrate on revising the measurements in trying to reduce the fluctuation for many minutes.

(3) Record the data

Let us consider the measurements of an electric circuit in which the current flowing in the circuit is measured as a function of the voltage of the power supply. Both of the current and voltage are measured with digital multi-meters. The data may be written down in the way shown in Fig. 7.1, as a list of the voltage values with corresponding current values. It is important first to record the physical quantities with their units at the top in the list (You may have some points deducted if you do not write down the units on the answer sheet). In this case, you do not need to make the measurements in the order of increasing voltage from the smallest values with some fixed step. As mentioned before, it is better to first roughly scan the required range with a large voltage step and measure the corresponding change in current roughly. Then, if needed, you can perform finer measurements only in the important voltage range where the change in current is rapid or where the phenomenon in question occurs. While, in this case, the data table is not in the order of ascending voltage, it will not be a problem points-wise.

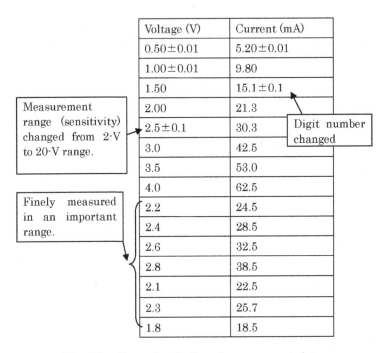

	Voltage (V)	Current (mA)
	0.50±0.01	5.20±0.01
	1.00±0.01	9.80
	1.50	15.1±0.1
Measurement range (sensitivity) changed from 2-V to 20-V range.	2.00	21.3
	2.5±0.1	30.3
	3.0	42.5
	3.5	53.0
	4.0	62.5
Finely measured in an important range.	2.2	24.5
	2.4	28.5
	2.6	32.5
	2.8	38.5
	2.1	22.5
	2.3	25.7
	1.8	18.5

Digit number changed

Fig. 7.1. Example of a list of measurement data.

When you change the measurement range of the digital multimeter, the unit may change from A to mA and/or the smallest digit may change in the display of the meter. You should be careful regarding such changes, i.e., recording the change in measurement error as shown in Fig. 7.1.

If possible, along with recording the measurement data in the list, it is recommended that you plot the data in a graph *during* the measurement. As shown in Fig. 7.2, the graph based on the data in Fig. 7.1, the horizontal axis is for the pre-set parameter (voltage in this case), and the vertical axis is the measurement result (current in this case). By drawing the graph, you can easily determine the specific range of the parameter where significant change occurs. In the example shown in Fig. 7.2, we can see that the gradient changes around 2.2 V. Therefore, a more precise measurement needs to be made only around this voltage value. In order to draw the graph, you need to find the maximum (and minimum) values on both axes.

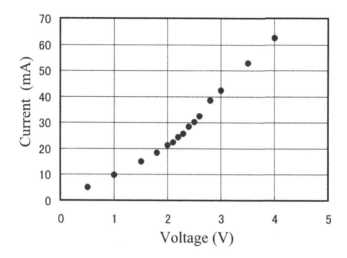

Fig. 7.2.   A graph of data in Fig. 7.1

For this, making a first quick measurement across the whole range of the parameter as mentioned before is very useful.

Of course, during experiments in which you cannot make such a preliminary quick measurement, you should change the preset parameter from the smallest value with a small step and perform the measurements steadily. At the experimental competition in IPhO2008 Vietnam, where the contestants were required to measure the current in an electric circuit every 30 seconds, they did not have enough time to plot the data in a graph during the measurement; they could draw the graph only after all of the measurements were completed.

You will need to take actions that suit the occasion.

(4) Measurements with a vernier

Some instruments with a vernier, such as a slide caliper and a protractor, are sometimes used in IPhO competitions. Here the fundamentals of measurement techniques with a vernier are explained using a slide caliper in the example (Fig. 7.3(a)).

The vernier has a scale which divides the length of nine divisions on the main scale into ten divisions. In other words, the minimum division a on the main scale is 1 mm and that on the vernier b is

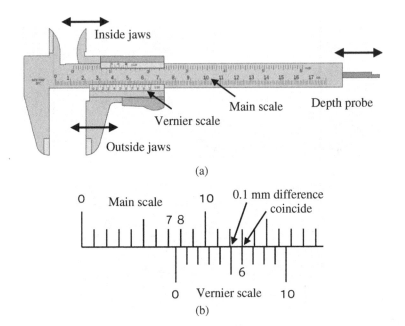

Fig. 7.3.   Caliper. (a) Names of parts. (b) Example of a measurement.

0.9 mm: $9a = 10b$. The difference in division between the main scale and the vernier is 0.1 mm, which is now the minimum division in the measurement. Therefore, the measurement error is ± 0.05 mm.

Figure 7.3(b) shows an example measurement. First you can do a rough measurement on the main scale at the position of origin on the vernier, indicating somewhere between 7 mm and 8 mm in this case. Next, look for the point where the lines on the main scale and on the vernier line up, and read the value at this point on the vernier scale. It is 6 for this example. As a result, the measurement value is 7.6 mm. Because, as mentioned above, the minimum division of the measurement using this caliper is 0.1 mm, the measurement error is 0.05 mm. Then, finally we get the measurement result of 7.60 ± 0.05 mm. This error is one order of magnitude smaller than that of a measurement with a simple ruler with a minimum division of 1 mm.

Exercise 7.1. Another type of caliper as shown in Fig. 7.4(a) is also widely used. The 39 mm in the main scale is divided into 20 divisions

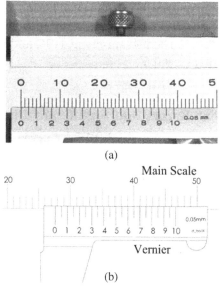

(a)

(b)

Fig. 7.4.

on the vernier. An example of measurement is shown in Fig. 7.4(b). Write down the measurement result with error for this case.

Answer The position at the origin on the vernier is located between 27 mm and 28 mm on the main scale. The position where the scale lines on the main scale and the vernier coincide is 6.5 on the vernier. Therefore, the measured value is 27.65 mm. Since the minimum division on the vernier is $39/20 = 1.95$ mm, which differs by 0.05 mm from two divisions on the main scale ($= 2$ mm), the minimum division in the measurement is 0.05 mm. Therefore, the measurement error is one half of the minimum division, or 0.025 mm (0.03 mm by rounding up). Finally, the measurement result is 27.65 ± 0.03 mm. (Do not record the result as 27.65 ± 0.025 mm: the third decimal place of the error value has no meaning because the second decimal place of the measured value already contains an error. The error should be written down with one digit only, as mentioned in the next section.)

A protractor used in an optical spectrometer has a vernier (similar to the experimental competition at IPhO2007 Iran). In this case, the minimum division on the main scale is $0.5°(= 30')$, while the vernier equally divides $29'$ on the main scale into 30 divisions. Then the minimum division in the measurement is $1'$, and its measurement error is half of that, i.e., $\pm 0.5'$.

7.2. Measurement Errors and Significant Figures

Since all results obtained in experiments contain some uncertainty (error), the measured result has no meaning if it is shown without the error component. Therefore, the concept of significant figures is important, which means how many digits in the measured value we can reply on. In other words, the concept of significant figures says that the value should be recorded in a way that only the last significant digit contains the error. For example, as mentioned before, when the length of an object is measured by a ruler with a minimum division of 1 mm, the error in the measurement is one half of the minimum division, i.e., ± 0.5 mm, meaning that the first place of the decimal contains the error and the second decimal place has no meaning. Then the measurement result should be recorded as 23.4 ± 0.5 mm, with the leftmost three digits for the significant figure, not as 23.40 ± 0.5 mm. In case of caliper measurements, the error is ± 0.05 mm, and the result should be written down as 23.40 ± 0.05 mm, meaning that the significant figure is four digits in this case and only the second decimal place contains the error.

For weight measurement with an instrument with an error in the 1 g digit, when the measured result is 0.134 kg, you can record the value as 1.34×10^{-1} kg to show explicitly that the significant figure is three digits. When you want to show the error together with the measured value, it should be written as $(1.34 \pm 0.01) \times 10^{-1}$ kg. In the case of measurements with the error in the 0.1 g digit, you can write down the result as $(1.340 \pm 0.001) \times 10^{-1}$ kg, showing explicitly that the significant figure is four digits. If you simply write down 300 g, it is not clear how many digits the significant figure is. When the significant figure is three digits, you should record it as 3.00×10^2 g. When it is four digits, 3.000×10^2 g is recommended. Be sure that only the last digit of the significant figure contains the error.

In the case of measurements with a digital multi-meter, when the measurement range is changed, you should be careful because the minimum digit changes automatically, and the error and significant figures will also change accordingly. For example, when you measure a voltage around 1.5 V, the result may be 1.4 V for the 20 V range, while it could be 1.38 V when the 2 V range is used. Be careful that the significant figure is now three digits and the error becomes smaller by one order of magnitude. The former result should be written as 1.4 ± 0.1 V, while the latter result should be 1.38 ± 0.01 V. Recall that the measurement error in a digital meter is given as one unit of the last digit in the display.

Exercise 7.2. How many digits is the significant figure for the following measured values?

(a) 0.00167 kg (b) 6400 g (c) 0.012300 kg (d) 100 mm

Answer (a) Three. (b) It is not clear by this value whether the significant figure is two or three or four digits. (c) Five. (d) It is not clear by this value whether the significant figure is one or two or three digits.

7.3. Statistical Errors

The measurement error so far mentioned is a reading error, originating from the discrete scales in the measurement instruments. This error depends on how small the sale division is in the instrument. Another type of errors is statistical error. We can minimize the statistical error by repeating the same measurements many times, or by reducing the ratio between the reading error and the measured value itself.

For example, let us consider an experiment in which the period of a pendulum is measured using a stopwatch. The results of ten measurements are e.g., 2.5 s, 2.7 s, 2.4 s, 2.6 s, 2.5 s, 2.3 s, 2.6 s, 2.6 s, 2.5 s, 2.7 s. (While the stopwatch displays the value down to the second decimal place in seconds, the order of 1/100 s. might be meaningless because the stopwatch was operated by hand; such manual operation will not ensure the precision of 1/100 s. Therefore, the significant figure may be down to the first decimal place.) The average value $\bar{T}$ of the n measured values $T_i (i = 1, \ldots, n)$ is defined by

$$\bar{T} = \frac{1}{n} \sum_{i=1}^{n} T_i. \tag{7.1}$$

Then the result is $\bar{T} = 2.540\,\text{s}$. Here we write the value to the third place of decimal (we will consider the significant figure later). The error ΔT is defined by

$$\Delta T = \sqrt{\frac{1}{n(n-1)} \sum_{i=1}^{n} (T_i - \bar{T})^2}, \tag{7.2}$$

or

$$\Delta T = \frac{1}{n} \sum_{i=1}^{n} |T_i - \bar{T}|. \tag{7.3}$$

This is an average of remainders $|T_i - \bar{T}|$ which means how much the i-th measured value T_i deviates from the average $\bar{T}$. The error calculated by Eq. (7.2) is called by the sample standard deviation. While the error ΔT calculated by Eq. (7.3) is usually larger than that in Eq. (7.2), both equations are available for calculation. Our pendulum measurement gives $\Delta T = 0.04\,\text{s}$ according to Eq. (7.2). Then, since we have an error in the second decimal position, the third decimal place has no meaning and so the measured result should be written as $T = 2.54 \pm 0.04\,\text{s}$ with considering the significant figure. It might be a tip for data analysis that you had better take largest number of digits available during the calculation and then round up the value into the significant figures properly at the end. This result looks reasonable because the error on the order of 0.1 s is inevitable for manual measurements with a stopwatch, as mentioned, but the final result contains an error smaller than 0.1 s thanks to the ten measurements.

However, you can easily imagine that a more precise measurement can be made by measuring the elapsed time for ten oscillations of the pendulum at once, instead of measuring the elapsed time of a single oscillation. The period can be obtained by dividing the elapsed time for the ten oscillations by ten. The elapsed times measured for the ten oscillations are 25.4 s, 25.3 s, 24.8 s, 25.9 s, 24.3 s, 25.0 s, 25.3 s, 25.1 s, 24.8 s, 25.9 s. Notice here that the significant figure is now three digits. By following the same procedure of analysis mentioned above, we obtain the average $\bar{T} = 2.518$ s. and the error $\Delta T = 0.015$ s. This means that the second decimal place contains the error. Then, finally we obtain the result $T = 2.52 \pm 0.02$ s. By comparing the previous result in which the single oscillation was measured, the measurement precision is now improved by a factor of two. The magnitude of error (about 0.1 s) at each operation of the stopwatch, which is inevitable for manual measurements, is now minimized with respect to the measured values themselves (around 25 s) by measuring the elapsed time for ten oscillations at once. In this way, we can reduce the statistical error by reducing the ratio (error)/(value). The present measurements (=10-times measurements of 10-periods) correspond to making 100 repeated measurements of a single oscillation.

This approach for measurements and data analysis are sometimes required in IPhO experimental competitions. An example of a plausible question is as follows: Measure the wavelength of a wave with an error smaller than 0.2 mm by using a ruler with a minimum division of 1 mm. (Actually similar question was at the experimental competition in IPhO2006 Singapore.) As mentioned several times before, since the reading error is one half of the minimum division, i.e., ±0.5 mm (which is larger than the required error), the contestants should recall the above-mentioned method to reduce the measurement error by repeating the measurements many times or by measuring the length of, e.g., ten wavelengths at once.

Exercise 7.3. (a) We repeated the measurements of the length of an object five times by using a caliper with the minimum division of 0.05 mm. The results are shown below in mm units. Calculate the average and error, and record the final result with error.

No. 1; 41.53,   No. 2; 41.49,   No. 3; 41.48,

No. 4; 41.51,   No. 5; 41.47

(b) We continued the same measurements 15 times more, and obtained the results below. By using all of the data (20 data points in total including those in (a)), calculate the average and the error. Write down the final result with error.

No. 6; 41.50,   No. 7; 41.49,   No. 8; 41.49,   No. 9; 41.52,

No. 10; 41.51,   No. 11; 41.51,   No. 12; 41.50,   No. 13; 41.51,

No. 14; 41.50,   No. 15; 41.49   No. 16; 41.51,   No. 17; 41.48,

No. 18; 41.49,   No. 19, 41.51,   No. 20; 41.50

Answer The average and error are calculated from Eqs. (7.1) and (7.2), respectively. (a) The average is 41.496, and the error is 0.0108. Since the second decimal place contains an error, the significant figure is four digits. Then, the final results is 41.50 ± 0.01 mm. The error is only slightly smaller than the reading error of each measurement (0.03 mm) because there are only five measurements. (b) The average is 41.499, and the error is 0.0033. Then, the result is 41.500 ± 0.003 mm, in which the error is smaller than in (a), because of the larger number of repeated measurements.

Exercise 7.4. (a) The data measured for the wavelength of a standing wave, taken five times with a ruler with a minimum division of 1 mm, are 23.5 mm, 23.7 mm, 24.0 mm, 23.3 mm 23.6 mm. Write down the final result with error.

(b) The data measured for ten wavelengths of the same standing wave, measured five times in the same way, are 234.5 mm, 235.0 mm, 235.5 mm, 234.6 mm 235.1 mm. Write down the final result with error.

Answer (a) According to Eqs. (7.1) and (7.2), the average is 23.62 mm, and the error is 0.12 mm. The final result is thus 23.6 ± 0.1 mm. The error here is smaller than the reading error at each measurement (which is one half of the minimum division (± 0.5 mm)) thanks to the five-time measurements. (b) The average is 23.494, and the error is 0.018, meaning that the second decimal place contains the error. Therefore, the final results is $23.49 \pm 0.02\,\mathrm{mm}$. These measurements are of much higher precision than in (a).

7.4. Errors in Indirect Measurements and Error Propagation

Let us consider a measurement in which, for example, by measuring the diameter of a circle with a ruler (its minimum division is 1 mm), and the area of the circle is calculated with error. If the diameter x is $x = 6.5 \pm 0.5\,\mathrm{mm}$, then the area y is calculated as $y = \frac{\pi}{4}x^2 = 33.18\,\mathrm{mm}^2$. We need to estimate the range of y, $y \pm \Delta y$, corresponding to the range of x, $x \pm \Delta x$, due to the error. By referring to Fig. 7.5, we can see how to estimate Δy;

$$\Delta y = |y(x + \Delta x) - y(x)|$$
$$= \left| \left(\frac{dy}{dx} \right) \cdot \Delta x \right|. \tag{7.4}$$

Therefore, since $dy/dx = \pi x/2 = 10.21$, $\Delta y = 10.21 \times 0.5 = 5.11\,\mathrm{mm}$, meaning that the first digit in y contains the error. Then, the final result of the area is $33 \pm 5\,\mathrm{mm}^2$. Since the first digit contains the error, the subsequent decimal digits in the value of y are not relevant.

This method can be extended to cases with more than two variables. For example, let us obtain the area of a rectangle by measuring the vertical and horizontal sides. The sides are measured with a ruler with a minimum division of 1 mm. The measured values

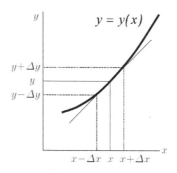

Fig. 7.5.

of the sides are, e.g., 8.2 mm and 17.5 mm. Then, the measurement results of sides are 8.2 ± 0.5 mm and 17.5 ± 0.5 mm. Then, the area, which is calculated as the product of the two sides, should contain some error.

In general, when a quantity y is a function of n variables $x_i (i = 1, 2, \ldots, n)$; $y(x_1, x_2, \ldots, x_n)$, and each of x_i has a measurements error Δx_i, the error Δy of y is obtained by the following equation, which is an extension of Eq. (7.4),

$$\Delta y = \sqrt{\left(\frac{\partial y}{\partial x_1} \Delta x_1\right)^2 + \left(\frac{\partial y}{\partial x_2} \Delta x_2\right)^2 + \cdots \left(\frac{\partial y}{\partial x_n} \Delta x_n\right)^2}. \quad (7.5)$$

The area y of the rectangle is a product of the two sides $x_1 (= 8.2 \pm 0.5$ mm$)$ and $x_2 (= 17.5 \pm 0.5$ mm$)$; $y = x_1 \cdot x_2 (= 143.5$ mm$^2)$. Next, by taking the partial derivatives, $\frac{\partial y}{\partial x_1} = x_2$, $\frac{\partial y}{\partial x_2} = x_1$, we obtain

$$\Delta y = \sqrt{(x_2 \cdot \Delta x_1)^2 + (x_1 \cdot \Delta x_2)^2}$$
$$= \sqrt{(17.5 \times 0.5)^2 + (8.2 \times 0.5)^2} = 9.66 \text{ mm}^2. \quad (7.6)$$

Since the significant figure of x_1 is two digits and that of x_2 is three digits, the significant figure of y should be two digits (see below). Then, the final result for the area is $y = 140 \pm 10$ mm^2, or $y = (1.4 \pm 0.1) \times 10^2$ mm^2 (to explicitly show the two digits of the significant figure).

Specifically, when it is written as $y = x_1^a \cdot x_2^b \cdots x_n^m$, according to Eq. (7.5), the relation between the error Δx_i of each x_i and that Δy of y is given by

$$\left|\frac{\Delta y}{y}\right| = \sqrt{\left(a\frac{\Delta x_1}{x_1}\right)^2 + \left(b\frac{\Delta x_2}{x_2}\right)^2 + \cdots \left(m\frac{\Delta x_n}{x_n}\right)^2}, \quad (7.5')$$

which is in a good form to remember.

Figure 7.6 shows why the significant figure of y in the above example is two digits. In general, the significant figure of a product or a quotient of the measured values should have the same digit number as the smallest significant figure digit number among the measured

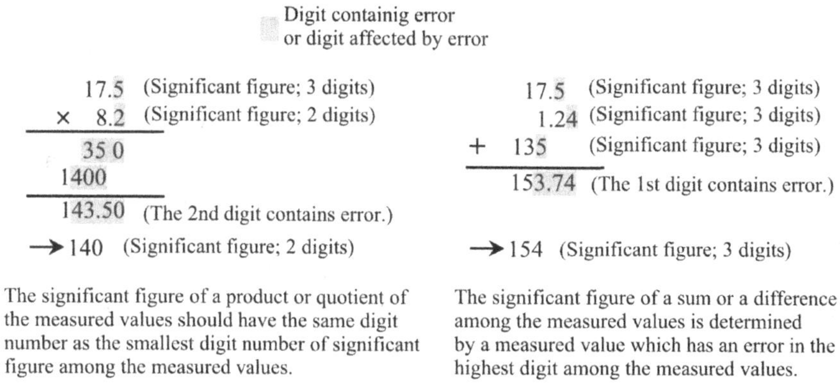

Digit containig error
or digit affected by error

17.5	(Significant figure; 3 digits)
× 8.2	(Significant figure; 2 digits)
35 0	
1400	
143.50	(The 2nd digit contains error.)
→ 140	(Significant figure; 2 digits)

17.5	(Significant figure; 3 digits)
1.24	(Significant figure; 3 digits)
+ 135	(Significant figure; 3 digits)
153.74	(The 1st digit contains error.)
→ 154	(Significant figure; 3 digits)

The significant figure of a product or quotient of the measured values should have the same digit number as the smallest digit number of significant figure among the measured values.

The significant figure of a sum or a difference among the measured values is determined by a measured value which has an error in the highest digit among the measured values.

Fig. 7.6. How to determine the significant figure of results with four arithmetical operations of the measured values.

values. The significant figure of a sum or a difference among the measured values is determined by the measured value which has an error in the highest digit among the measured values. But there are exceptions as seen in Exercise 7.6. In order to determine the significant figure of the final result, you should be always careful to notice which digit contains the measurement error in each value.

Exercise 7.5. Write down the results of the following calculations with the appropriate significant figures.

(a) 53×27 (b) 37.9×75 (c) $41.53 \div 3.8$

Answer (a) Because the significant figures of both numbers are two digits, that of the product should be also two digits; 1400. (b) The significant figure of the product should be two digits because those of the two numbers are three and two digits; 2800. (c) The significant figure of the quotient should be two digits because those of the two numbers are four and two digits; 11.

Exercise 7.6 Write down the results of following calculations with two measured values (with error)

(a) $(8.3 \pm 0.5) \times (25.2 \pm 0.5)$ (b) $(2.55 \pm 0.05) \times (23.2 \pm 0.5)$

(c) $(8.3 \pm 0.5) + (25.2 \pm 0.5)$ (d) $(2.55 \pm 0.05) + (23.2 \pm 0.5)$

Answer The errors of the products and sums can be calculated from Eq. (7.5). (a) and (b) are in a form of $y = x_1 \cdot x_2$, while (c) and (d) are in a form of $y = x_1 + x_2$. (a) The significant figure of the product should be two digits; 210 ± 10. (b) The product is 59.16, and the error is 1.73. Since an error is included in the first place, the result is 59 ± 2. Notice that the significant figure should be three digits because those of the two numbers are three digits in this case. However, since the first place of the product already contains an error, the first decimal digit is not relevant. This is an exceptional example of the rule mentioned in Fig. 7.6. (c) The sum is 33.5, and the error is 0.71. Since the sum contains the error only in the first decimal place, the result is 33.5 ± 0.7. Be aware that the result has three significant figures although one of the numbers has only two significant figures. (d) The sum is 25.75, and the error is 0.502. Since the first decimal place contains the error, the result is 25.8 ± 0.5. Notice that when the error of one of the two values is much larger than that of the other, such as in this example, the error of the final result is governed by the lager error.

7.5. Best-fit to a Linear Function

In junior-high schools, we learn Hooke's law, which states that the extension x of a spring is proportional to the applied force f, written by the equation $f = kx$, where the proportional coefficient k is called the spring constant. Let us consider an experiment to obtain the k value of a given spring. The measurement data is shown in Table 7.1. The extension of the spring was measured using a ruler with a minimum division of $1 \, \text{mm}$. For simplicity, the errors in the masses of the weights and the acceleration due to gravity are assumed to be negligibly small.

First, the data should be plotted on a graph. Each data point should be accompanied by an error bar to show the magnitude of the error. In this case, the error bars are only in the extension of the spring. Figure 7.7 shows the applied force f (calculated from the masses of the weights) on the vertical axis, and the resulting extension x of the spring on the horizontal axis. This plot is reversed

Table 7.1.

Weight (g)	Force f (N)	Extension x (mm)
0	0	0
10	0.098	2.0 ± 0.5
20	0.20	7.2
30	0.29	10.3
40	0.39	11.8
50	0.49	13.7
60	0.59	18.2
70	0.69	20.7
80	0.78	23.7
90	0.88	26.3
100	0.98	30.4

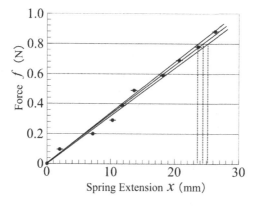

Fig. 7.7.

on axes compared with the usual method in which the preset parameter (f for this case) should be on the horizontal axis. But this special type of plot is for a purpose that we can directly obtain the proportional coefficient k in the Hooke's law $f = kx$ from this graph. The law says that the data should fit on a straight line passing through the origin. The gradient of the line should be the spring constant k.

Since the contestants in IPhO or in some other competitions, are not allowed to use a computer, the fitting of a straight line to the measured data should be done by eye measurement. By considering the data scattering and error bars on the graph, together with the restriction (in this case, the restriction is that the line should ' pass through the origin), three lines should be drawn, one with the maximum gradient, a line with the minimum gradient, and finally the most plausible line which should be between the two other lines. The error in the spring constant k is deduced from the upper and lower lines. Figure 7 gives the most plausible line with $k = 0.8/(24.4 \times 10^{-3}) = 32.8\,\text{N/m}$. The upper and lower lines give a range $k = 31.8 \sim 33.8\,\text{N/m}$. Therefore, the error is roughly $\pm 1.0\,\text{N/m}$, meaning that the first place in the value of k contains an error component and the significant figure is down to the first place. Then, we obtain the final result of this experiment as $k = 33 \pm 1\,\text{N/m}$. Some students may draw the upper and lower lines differently, giving a larger error for k. This would be OK, however, if the upper and lower lines are plausible by considering the scattering of data points and their error bars. Even if each measured value of the extension x has three digits of significant figure, the final result have only two digits of significant figure because the data points are so scattered in this case.

Since, in this example, it is obvious (theoretical) that the fitting lines should pass through the origin, this restriction is used for the fitting procedure. But, in some cases, the pre-set variable x and the measurement result y have a relation such as $y = ax + b$, from which the values of both a and b need to be determined with their errors. In such cases, the three lines should be determined by changing not only the gradient, but also the intercept so that the values and their errors of a and b can be obtained (see Exercise 7.7).

In some experimental competitions in IPhO, furthermore, the measurement result y is related with the pre-set variable x by an equation such as $y = ax^2$, and the contestants are asked to obtain the coefficient a together with its error experimentally. In this case, the graph should not be drawn in x vs. y, but in x^2 vs. y by calculating the values of x^2 in advance (with their errors, if necessary). Then you can obtain the value and error of a by fitting the straight line of the data x^2 vs. y on the graph. In IPhO, it is never required to perform data fitting with nonlinear curves. However, by making some calculation in advance from the measured data, the quantities plotted on the horizontal and vertical axes can be changed to make the linear fitting possible. Such technique is frequently used in IPhO. In fact, in IPhO2008 Vietnam, the contestants were required, from the measured quantities, to find the quantities for the horizontal and vertical axes on a graph to make data analysis possible by fitting a straight line.

Exercise 7.7. (A part of the experimental competition in IPhO2005 Spain) The voltage V across a miniature bulb was measured by flowing the current I through it, and the resistance $R(=V/I)$ of the bulb was obtained as a function of I. The data are shown in Table 7.2. It is shown that the resistance increases with the current because of the heating of the filament in the bulb. From the data, obtain the resistance value (with its error) of the bulb at room temperature (i.e., without current flowing).

Table 7.2.

Current I (mA)	Voltage V (mV)	Resistance $R(\Omega)$
1.87 ± 0.01	21.9 ± 0.1	11.7 ± 0.01
2.58	30.5	11.8
2.95	34.9	11.8
3.12	37.0	11.9
3.37	40.1	11.9
3.60	43.0	11.9
3.97	47.6	12.0
4.24	51.1	12.1
4.56	55.3	12.1
4.79	58.3	12.2
5.02	61.3	12.2
5.33	65.5	12.3
5.47	67.5	12.3
5.88	73.0	12.4
6.42	80.9	12.6
6.73	85.6	12.7
6.96	89.0	12.8
7.36	95.1	12.9
8.38	112 ± 1	13.4
9.37	130	13.9
11.7 ± 0.1	182	15.6

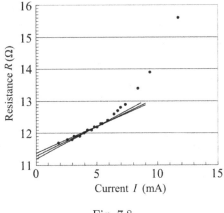

Fig. 7.8.

Answer The resistance is plotted as a function of the current as shown in Fig. 7.8. It clearly shows that the resistance significantly increases with the current (in a nonlinear manner), while the resistance is nearly proportional to the current in the range of low currents. Therefore, we can fit the data less than 6 mA with a straight line. As shown in Fig. 7.8, by considering the data scattering, we can draw three plausible straight lines (in this case, the error bars are so small that we only have to consider the data scattering). The intercept at the vertical axis gives the required value of the resistance value at zero current. By considering the scattering of the intercept, the result is $R = 11.3 \pm 0.1\,\Omega$. The filament in the bulb is kept at room temperature without current flowing. But we cannot directly measure the resistance without current flowing through the filament. This kind of data analysis is called 'extrapolation' in which the values outside of the measurement range are estimated from the data available by assuming some function (a linear function in this case).

7.6. Best-fit to a Logarithmic Function

Let us consider an experiment in which a translucent glass with a thickness d is irradiated by light with intensity I_0, and the intensity I of the transmitted light is measured behind the glass. The relation

Table 7.3.

Glass Thickness d (mm)	2.5 ± 0.5	6.0	8.0	10.5	12.0
Intensity I (V)	4.34 ± 0.1	1.89	0.822 ± 0.005	0.365	0.195
$\ln(I)$		1.47	0.637	-0.196	-1.01	-1.63

between d and I is given by

$$I(d) = I_0 \exp\left(-\frac{d}{\lambda}\right), \tag{7.7}$$

where the constant λ is called the extinction length in the glass. The intensity I of the transmitted light is measured using glass of different thicknesses. The purpose of the experiment is to obtain the extinction length in the glass from the data. The data is shown in Table 7.3. The intensity of the transmitted light was measured with a photodiode which outputted a voltage proportional to the light intensity. The thickness of the glass was measured with a ruler having a minimum division of 1 mm.

Since we cannot do the data fitting by a straight line directly on a graph of I vs. d, we have to consider some data conversion. By taking the natural logarithm of both sides in Eq. (7.7), we obtain

$$\ln(I(d)) = -\frac{d}{\lambda} + \ln(I_0). \tag{7.8}$$

This means the data plotted on a graph with d on the horizontal axis and $\ln(I)$ on the vertical axis, should be on a straight line. The inverse of its gradient is λ. Therefore, we first calculate $\ln(I)$ from the measured values of I, which are also shown in Table 7.3. The graph is shown in Fig. 7.9 (If you use a semi-log graph, you do not need to calculate $\ln(I)$. But it seems that regular section papers are always used in IPhO, with calculating the logarithmic values in advance by a calculator). As mentioned before, we can now fit to straight lines such that three lines should be drawn by eye measurement while considering the data scattering and error bars. From the most plausible line, we can obtain the gradient and then its inverse; $(11.7 - 2.8)/3 = 2.97$. The upper and lower straight lines similarly gives the two values of λ, 2.80 and 3.10, meaning the error

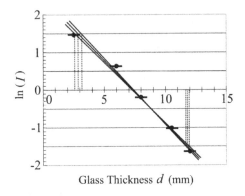

Fig. 7.9.

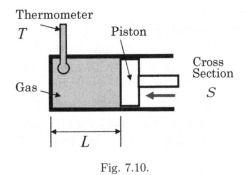

Fig. 7.10.

is ± 0.15. As a result, we obtain $\lambda = 3.0 \pm 0.2\,\mathrm{mm}$ (because the first decimal place contains an error).

Exercise 7.8. Let us consider an experiment using a cylinder with a piston with thermal insulating walls as shown in Fig. 7.10. By pressing the gas confined in the cylinder with the piston, the volume V and temperature T of the gas are measured. In this case, the relation between V and T

$$T \cdot V^{\gamma - 1} = \text{constant},$$

is known (Poisson's law), where γ is a constant value called "heat capacity ratio". In this experiment, the length L of the confined area of the cylinder was measured instead of the volume of the gas (because the cross section S of the cylinder is constant). The data

Table 7.4.

L (mm)	300.0 ± 0.5	275.0	250.0	225.0	200.0	175.0
T (°C)	25.2±0.5	36.1	51.7	62.8	83.2	97.5

are shown in Table 7.4. From these data, obtain the γ value in the above equation (together with its error).

L was measured with a ruler having a minimum division of 1 mm, while the T was measured with a stem thermometer having a minimum division of 1°C.

Answer Since the volume of the gas is $V = SL$, by taking the natural logarithm of both sides in the above equation, we get

$$\ln T = -(\gamma - 1) \cdot \ln L + C, \tag{7.9}$$

with an appropriate constant C. Therefore we can plot the data on the graph in which $\ln T$ and $\ln L$ are on the vertical and horizontal axes, respectively (Do not forget to convert the measured temperature in °C into the absolute temperature in K). Then, by fitting straight lines to the data on the graph, we can obtain the γ value from the gradient. The graph is shown in Fig. 7.11 (the error bars on the data points are too small to see). As mentioned several times before, three straight lines should be drawn. From the line,

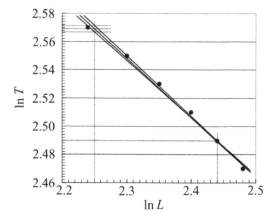

Fig. 7.11.

we can get the gradient $(2.569 - 2.490)/(2.44 - 2.25) = 0.416$. The upper and lower lines similarly give the gradients, 0.402 and 0.432, respectively, with the error ± 0.015. Thus, it is clear that the second decimal place contains an error. Finally, we get $\gamma = 1.42 \pm 0.02$.

Incidentally, the γ is a ratio between the specific heat at constant pressure and that at constant volume. In case of an ideal gas, it is known theoretically $\gamma = 5/3 = 1.67$ for monatomic molecule gas, while $\gamma = 7/5 = 1.40$ for diatomic molecule gas.

7.7. Summary

Important points for experimental measurements, data recording and analysis, and error evaluation are summarized as "the 10 articles" below.

① In the case of measurements with instruments, such as a ruler and a stem thermometer on which the scale is marked with fixed divisions, the reading error (uncertainty) is one half the minimum increment. In the case of measurements with instruments having a digital display of values, such as a digital multi-meter, the reading error is one increment of the last digit.

② When the measured value on such an instrument fluctuates due to noise or other unknown reasons (regardless of whether the instrument is analog or digital), you should read the central value in the fluctuation and put the fluctuating range as the reading error. You should not adhere to the scale division or the last digit in the display of the measurement instrument for the reading error.

③ When the measurement range (sensitivity) in the instrument is changed, be careful to also change the reading error and significant figure of the measured value.

④ The error should be, in principle, only one digit.

⑤ The significant figure is written in a way that the value contains the error only in the last digit.

⑥ When the measured values are scattered from measurement to measurement, you should repeat the measurement several times, and regard <u>the standard deviation calculated from these data</u> <u>as the measurement error.</u> (But you should not spend too much time on the repeated measurements — keep in mind the time allotted.)

⑦ <u>By repeating the same measurement several times, you can</u> <u>reduce the measurement error (statistical error)</u> and improve the precision. (But you should not improve the precision more than that you need.)

⑧ In addition to repeating the same measurement, <u>you can reduce</u> <u>the measurement error by reducing the ratio of the reading error</u> <u>to the measured value.</u> For example, you can measure the elapsed time for ten oscillations of a pendulum, rather than that of a single oscillation.

⑨ The calculation of error should be clearly recorded on your answer sheets <u>by writing down the equations</u> you have used for error estimation, together with the numerical values. The errors you write down on your answer sheet are meaningless without showing any bases for the error estimation.

⑩ You can estimate the measurement errors not only through numerical calculations, but also <u>by using best-fit methods on a</u> <u>graph.</u> Three straight lines should be clearly drawn on the graph, which is an important basis for evaluating the error.

Chapter 8

Practical Exercises

In this chapter, two examples of experimental exercises are shown (Practical Exercises 1 and 2), both of which were actually used for Physics Challenge (the domestic competition for IPhO in Japan).

Practical Exercise 1 Experiments of Boyle's Law and Charles' Law, and a measurement of the atmospheric pressure by using an cylinder (from the First Challenge).

Problem 8.1. Confirming Boyle's law

50 ml air is confined in a cylinder which is connected with a pressure gauge as shown in Fig. 8.1. Boyle's law says that the product of pressure P and volume V of the confined gas should be constant if the temperature remains constant: $P \cdot V = c$ (constant). By confirming this law experimentally, obtain the value c with its error. The minimum division in the scale on the cylinder is 2 ml, and that on the pressure gauge is 5 kPa.

Answer By pushing or pulling the piston to change the pressure and volume of the air confined in the cylinder, the data in Table 8.1 were obtained.

The values of the product $c(= P \cdot V)$ are calculated by considering the significant figure. The error of each product are calculated by $\Delta c = \sqrt{(P \cdot \Delta V)^2 + (V \cdot \Delta P)^2}$, where $\Delta V = 1$ ml and $\Delta P = 3$ kPa. A graph in Fig. 8.2 is drawn in which the horizontal axis is the pressure P and the vertical axis is the product c. By drawing three lines

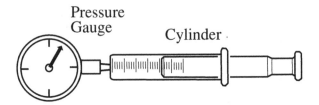

Fig. 8.1.

Table 8.1.

Pressure P (kPa)	150 ± 3	140 ± 3	130	120	102	90	80	70	65
Volume V (ml)	34 ± 1	3 ± 1	39	42	51	59	68	79	85
$c(= P \cdot V)$ (Pa · l)	5100 ± 200	5000 ± 200	5100 ± 200	5000 ± 200	5200 ± 200	5300 ± 200	5400 ± 200	5500 ± 200	5500 ± 300

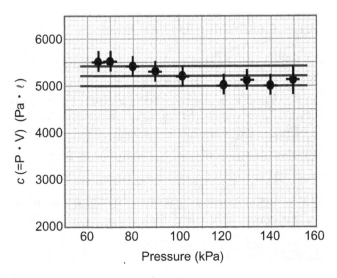

Fig. 8.2.

in the graph, the average of c and its error are obtained: $c = 5200 \pm 200 \,\text{Pa} \cdot \text{l}$.

Be noticed that the significant figure is two digits, and the final error is not smaller than the error of each measurement of c

probably because of some systematic error (such as leakage of air, change in temperature, etc); the data in the figure actually change systematically, rather than randomly.

Of course, you can calculate the average and the error of c numerically from the nine data of $c_i (i = 1 \sim 9)$ in the table by using the equations;

$$\bar{c} = \frac{1}{9} \sum_{i=1}^{9} c_i = 5200, \quad \Delta c = \sqrt{\frac{1}{9 \cdot (9-1)} \sum_{i=1}^{9} (c_i - \bar{c})^2} = 70.$$

Therefore $\underline{c = 5200 \pm 100 \text{ Pa} \cdot \text{l}}$.

Another Solution From the data in Table 8.1, we can calculate the values of $1/P$, and draw a graph of $1/P$ vs. V as shown in Fig. 8.3. By fitting a straight line $V = c/P$ to the data to obtain the gradient, we can get the value of c with its error.

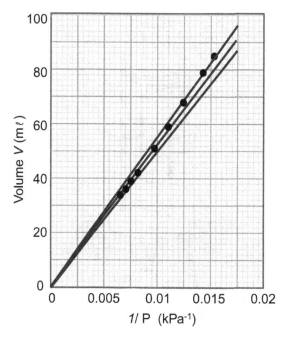

Fig. 8.3.

Ref According to the ideal gas equation $P \cdot V = n \cdot R \cdot T$, where n is the amount of the gas in mol, R is the gas constant, and T is the temperature of the gas, the constant c can be calculated. Since $R = 8.31 \, \text{J/K/mol}$, $T = 290 \, \text{K}$ and $n = 50/(22.4 \times 10^3) = 2.24 \times 10^{-3} \, \text{mol}$,

$$c = n \cdot R \cdot T = 5.4 \, J = 5400 \, Pa \cdot l,$$

which is roughly consistent with our experimental result within the error.

Problem 8.2. Confirming Charles' law

Some amount of air is confined in a cylinder of which end is closed by a rubber plug. The whole cylinder is immersed in water in a beaker on a gas stove as shown in Fig. 8.4 to change the temperature of the air inside. Charles' law says that the volume V of the confined gas should be proportional to the temperature T, which is $V = k \cdot T$, with a proportional constant k. By confirming this law experimentally,

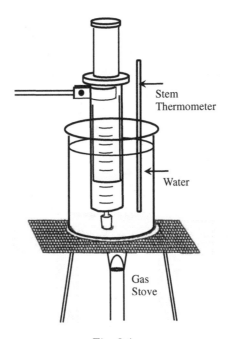

Fig. 8.4.

obtain the value k with its error. The minimum division in the scale on the cylinder is 2 ml, and that on the stem thermometer is 1°C.

Answer By slowly heating the water, we measured the volume of the confined air as a function of temperature. The data below were obtained.

According to the data in Table 8.2, we have drawn a graph as shown in Fig. 8.5. By considering the data scattering and error bars, we drew three straight lines. Then, the gradient of the central line gives the value of k and the upper and lower lines give its error. By considering the significant figures of the measured data, we finally obtain the result, $k \pm \Delta k = 0.18 \pm 0.01$ ml/K.

Table 8.2.

Temperature (°C)	6 ± 0.5	13 ± 0.5	33 ± 0.5	42 ± 0.5	56 ± 0.5	70 ± 0.5
Temperature T (K)	279 ± 0.5	286 ± 0.5	306 ± 0.5	315 ± 0.5	329 ± 0.5	343 ± 0.5
Volume V (ml)	49 ± 1	50 ± 1	54 ± 1	56 ± 1	58 ± 1	60 ± 1

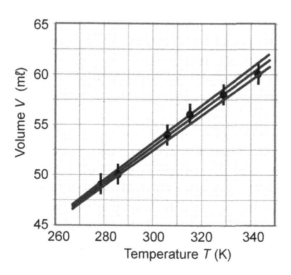

Fig. 8.5.

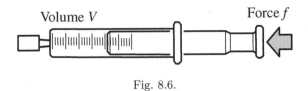

Fig. 8.6.

Ref According to the ideal gas equation $P \cdot V = n \cdot R \cdot T$, $k = nR/P$. We can estimate the amount of air $n = 48/(22.4 \times 10^3) = 2.14 \times 10^{-3}$ mol. The pressure is $P = 1$ atom $= 101.3$ kPa, and $R = 8.31$ J/K/mol. Then k is calculated to be $k = 0.176$ ml/K, which agrees well with our experimental result.

Problem 8.3.   Measuring the atmospheric pressure

By measuring the volume of air confined in a cylinder and the force pushing the piston (see Fig. 8.6), obtain the atmospheric pressure.

Answer After confining the air of 100 ml inside the cylinder by closing its tip with a rubber plug, the cylinder was put on a platform scale as shown in Fig. 8.7. With pushing down the cylinder, we measured the volume V (ml) of the air inside the cylinder as a function of the applied force f (kgf) which was measured by the platform scale simultaneously. The data are summarized in the first two lines of Table 8.3. The unit of force is changed into Newton by multiplying 9.81 cm/s^2, which is listed in the third line of Table 8.3.

The piston is pushed from outside not only by the force f, but also by the atmospheric pressure P_0. Therefore, the total force pushing the piston from outside is $f + P_0 \cdot S$, where S is the cross section of the cylinder. On the other hand, the piston is also pushed from inside in the opposite direction, i.e., by the air confined. The force is $P \cdot S$, where P is the pressure of the confined air. Since the forces acting on the both sides of the piston is balanced, we can get

$$f + P_0 \cdot S = P \cdot S. \tag{8.1}$$

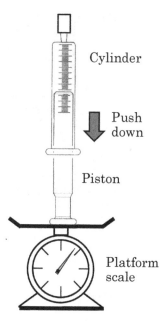

Cylinder

Push down

Piston

Platform scale

Fig. 8.7.

Table 8.3.

Volume V (ml)	100 ± 1	90	80	70	60	50	44
Force f (kgf)	0	1.1 ± 0.1	2.5	4.0 ± 0.2	6.3	9.3 ± 0.3	12
Force f (N)	0	11 ± 1	25	39 ± 2	62	91 ± 3	118
Inverse of Vol. $1/V$ (ml^{-1})	0.0100 ± 0.0002	0.011	0.013	0.014	0.017	0.020	0.023

On the other hand, by recalling Boyle's Law $P \cdot V = c$ (constant), we can eliminate P from Eq. (8.1), and get the relation

$$f = cS/V - P_0 \cdot S. \tag{8.2}$$

Therefore, when we draw a graph in which the vertical axis is f and the horizontal axis is $1/V$, the intercept at the vertical axis gives the value of $-P_0 \cdot S$. Then we can get the atmospheric pressure P_0 because we know the cross section S of the cylinder.

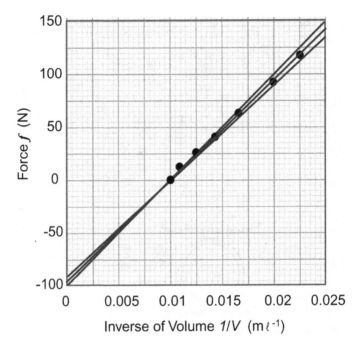

Fig. 8.8.

The inverse of volume is calculated with the error. Then, we have made the graph as shown in Fig. 8.8. The error bars are too small to see. When we fitted three straight lines to the data, we imposed a restriction that the lines should pass through a point of $f = 0(\mathrm{N})$ at $1/V = 0.01\,\mathrm{ml}^{-1}$ (which is the initial situation that the air of $100\,\mathrm{ml}$ at the atmospheric pressure is confined in the cylinder).

The intercept of the fitted lines at the vertical axis give

$$P_0 \cdot S = 95 \pm 5\mathrm{N}. \tag{8.3}$$

The cross section S of the cylinder was obtained by measuring the distance l between two scale lines showing $V = S \cdot \mathrm{l}$; $S = V/\mathrm{l} = 9.4 \pm 0.2\,\mathrm{cm}^2$.

Therefore, by combing Eq. (8.3), we get $P_0 = 100 \pm 6\,\mathrm{kPa}$. This result agrees nicely, within the error, with the known value of the atmospheric pressure, $101.3\,\mathrm{kPa}$.

Practical Exercise 2 Measuring Planck's constant (from the Second Challenge)

Purpose of Experiments

In Experiment 1, we will measure the wavelength λ of light emitted from light-emitting diodes (LED) by using interference phenomenon with a diffraction grating. From the λ, we will calculate the frequency ν of the light by a relation

$$\lambda = \frac{c}{\nu}, \tag{8.1}$$

where c is the speed of light. In Experiment 2, we will measure the energy E of the light. Then, by using the Einstein relation between E and ν,

$$E = h\nu, \tag{8.2}$$

together with the value of ν obtained in Experiment 1, we will be able to get the value of Planck's constant h. Throughout the experiments, we will understand the wave-particle duality of light, a basis of quantum physics.

Instruments and Parts used

(1) A simple spectroscope (Fig. 8.9)

This instrument is used for measuring the wavelength of light by using interference phenomenon produced by a diffraction grating. A rectangular box shown in Fig. 8.9 has a slit ①, through which the light under investigation comes in, and on the opposite end of the box, a window ② with the grating stuck on, through which an observer look into the box. The observer will see a spectrum with rainbow color on an inner wall next to the slit if you light through the slit ① with room light or a miniature bulb. You can measure the distance between the slit ① and the aimed color in the spectrum by using markers (thin openings) on a sliding plate ③. Move the sliding plate ③ with your hand to adjust the marker to the aimed color in the spectrum. And then measure the distance between the slit and the marker with a ruler on the outside of the box.

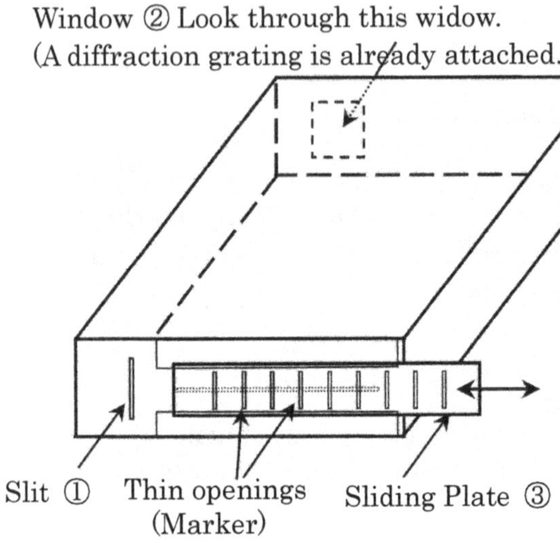

Window ② Look through this widow.
(A diffraction grating is already attached.)
Slit ①
Thin openings
(Marker)
Sliding Plate ③

Fig. 8.9. A simple spectroscope.

Clip terminal
Lead
Electrode
poles
LED(red)
LED(green)
Knob for
voltage
control
0 V
6 V
LED(blue)
Mini bulb

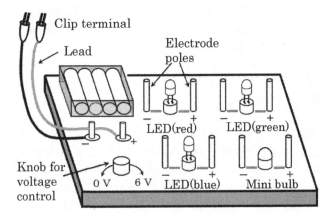

Fig. 8.10. Lighting board for LED and bulb.

Cautions

Be careful not to damage your eye and glasses when you look into the box through Window ② .

(2) Lighting board for LEDs and a miniature bulb (Fig. 8.10)

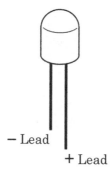

Fig. 8.11. LED.

The power for LEDs and a miniature bulb is supplied by four batteries in series with the maximum voltage of 6 V. The voltage can be controlled by a knob in a range 0~6 V as shown in Fig. 8.10. The clip terminals (red: positive, black: negative) of the batteries should be connected properly to the electrode poles of a LED or a miniature bulb in order to light them. In the case of LEDs, the polarity is important: Be sure that a longer lead of a LED is for positive and a shorter one for negative, as shown in Fig. 8.11. When connecting the clips, be careful not to connect positive to negative or vice versa.

Cautions

Be careful that LEDs can burst if the applied voltage is too high. The voltage should be always applied from zero volt and be increased slowly. LEDs can shine very brightly when the voltage is high enough. Do not look at a brightly illuminated LED for a long time in order to protect your eyes. Also, since the extreme brightness shortens the life of the LED, do not leave it in this condition for a long time.

Experiment 1

We will measure the wavelengths of red, green, and blue light emitted from LEDs, by using interference phenomenon produced by a diffraction grating. When you irradiate a diffraction grating with a laser beam as shown in Fig. 8.12, you will get diffraction spots on the screen behind. The grating has a few hundred evenly-spaced parallel

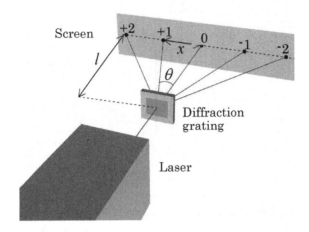

Fig. 8.12.   Interference experiment with a diffraction grating.

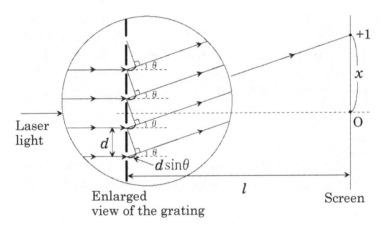

Fig. 8.13.   Rays of light diffracted by the grating.

fine grooves per mm. Light that falls in the grooves is scattered in various directions, and interferes with each other to form interference spots (diffraction spots) on the screen. Since the space d between the grooves on the grating is much shorter than the distance l between the grating and the screen, the rays of light from the grooves can be regarded parallel to each other, as shown in Fig. 8.13.

A diffraction spot is at a point located at x from the center of the screen, if the path difference between the two rays from the adjacent grooves to the point on the screen is a multiple of the wavelength λ of light. Since the path difference is $d \sin \theta$, this condition is written by

$$d \sin \theta = m\lambda \quad (m = 0, 1, 2, \ldots). \tag{8.3}$$

Substituting $\sin \theta = \frac{x}{\sqrt{l^2 + x^2}}$ into this equation gives

$$\frac{xd}{\sqrt{l^2 + x^2}} = m\lambda \quad (m = 0, 1, 2, \ldots). \tag{8.4}$$

On the same principle, we can also get a diffraction pattern by the simple spectroscope (Fig. 8.9) as shown in Fig. 8.14, in which the light under investigation is not purely monochromatic so that we will obtain colorful spectrum; since the x in Eq. (8.4) depends on the wavelength of light, different color appears with different x. Our grating has the spacing between the grooves $d = 2.00 \times 10^{-6}$ m. The condition of Eq. (8.4) for constructive interference of diffracted light

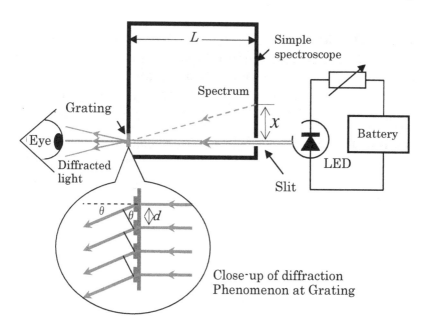

Fig. 8.14.   Measurement with the simple spectroscope.

for our spectroscope is written by

$$\frac{xd}{\sqrt{L^2 + x^2}} = m\lambda, \tag{8.5}$$

where L is the distance between the grating and the inner wall on which the spectrum is formed ($L = 220\,\text{mm}$). In our experiments, we always observed the first-order diffracted light only ($m = 1$).

Question 1 By introducing the light from each LED (red, green, and blue) into the simple spectroscope through the slit, measure the distance x for each color on the wall of the spectroscope. By moving the sliding plate ③ in Fig. 8.9 to bring a marker to the center of a bright line in the spectrum, you can measure the distance x between the marker and the slit from the outside of the spectroscope box by a ruler (sample data is shown in Table 8.4). If the bright line in the spectrum has some width so that it is difficult to determine the center position of the line, you can regard such uncertainty as the measurement error Δx. Based on the results, calculate the wavelength λ of light from each LED.

Question 2 Calculate the measurement error $\Delta\lambda$ of the wavelength of light from each LED, and record the final result of each wavelength with the error with the proper significant figure. You can assume for simplicity that $L\,(= 220\,\text{mm})$ and $d\,(= 2.00 \times 10^{-6}\,\text{m})$ in Eq. (8.5) contain no error.

Question 3 Calculate the frequency ν of light (together with the error) by combining Eq. (8.1) and the wavelength of each LED light obtained above. Here you can assume that the speed of light $c\,(= 3.00 \times 10^8\,\text{m/s})$ contains no error.

Table 8.4.

	Distance $x \pm \Delta x$ (m)
Red LED	$(7.4 \pm 0.1) \times 10^{-2}$
Green LED	$(6.3 \pm 0.1) \times 10^{-2}$
Blue LED	$(5.6 \pm 0.3) \times 10^{-2}$

Experiment 2

In this experiment, we will measure the current-voltage characteristic curve of each LED to show that the light of its frequency ν has energy of $h\nu$, and calculate Planck's constant h using the wavelengths obtained in Experiment 1.

The LED is made of materials called "semiconductors". In terms of electric current conductivity, materials around us are categorized roughly into three types. The first type is metal through which the electric current flows easily. The second type is insulator in which the current hardly flows. And the third one is semiconductor which has characters between metal and insulator. The electric current is in general a flow of electrons which can be found in any atoms. But the three types of materials are different from each other in the energy states of electrons and the amount of electrons that can move freely to create a current. The energy states of electrons in semiconductors can be schematically shown in a simple diagram in Fig. 8.15.

Most electrons contained in atoms that make up a crystal reside in lower energy states called "valence bands". The higher energy states are called "conduction bands". Between the valence and conduction bands, there is a forbidden energy gap in which no electrons can stay. Electrons in the conduction bands can flow through the material to contribute current, while electrons in the valence band cannot.

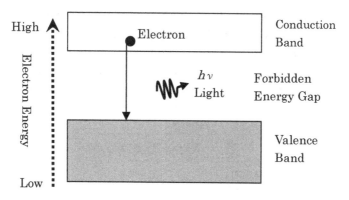

Fig. 8.15.   Energy states of electrons in semiconductors.

In LED, electrons in the valence band are pumped up into the conduction bands by using the energy from a battery. Then, when such electrons fall back to the valence band, the energy which equals the energy difference between the conduction and valence bands (i.e., the width of the forbidden energy gap), is released as light emission. The width of the forbidden energy gap is different from one semiconductor to another. Therefore, we can change the color of light, which corresponds to the wavelength or frequency of light, by tuning the gap energy of the LED semiconductor.

The energy of those electrons which are pumped up by a battery is $E = eV$, where V is the voltage of the battery and e is the elementary charge. When this energy is larger than eV_0, the minimum amount of energy required to leap over the forbidden gap, or in other word, when a voltage larger than V_0 is applied, the current begins to flow through the LED and to emit light. As mentioned before, the LED emits light when electrons fall across the forbidden gap from the conduction band to the valence band, the energy of light emitted is given by

$$E = h\nu = eV_0. \tag{8.6}$$

By connecting two digital multi-meters (one is as a direct current meter, and another is as a direct volt meter) with the Lighting board (Fig. 8.10) using some additional leads, measure the current flowing through a LED as a function of the voltage across it.

Question 4 By gradually increasing the voltage applied to a LED, find the threshold voltage V_0 at which the LED starts to emit light (The sample data are shown in Table 8.5). Because it is difficult to judge whether the LED lightens by naked eyes, you will have uncertainty in determining V_0. Then, you can regard such uncertainty

Table 8.5.

	Voltage $V_0 \pm \Delta V_0$ (V)
Red LED	1.6 ± 0.1
Green LED	1.8 ± 0.1
Blue LED	2.6 ± 0.1

as the measurement error ΔV_0. Next, calculate the energy of light $E = eV_0$ together with the error. Here you can assume that the elementary charge $e = 1.60 \times 10^{-19}$ C contains no error for simplicity.

Question 5 From the results including Experiment 1, draw a graph in which the vertical axis is E and the horizontal axis is ν. Do not forget to put error bars for each data point. Then, by fitting the data with Eq. (8.6), obtain the value of Planck's constant h, together with the error. Do not forget to include the unit.

Question 6 For each LED, measure the current flowing through as a function of the voltage (The sample data are shown in Table 8.6). Do not let a current larger than 20 mA flow. You should properly change the measurement range in the digital multi-meter; 2 V or 20 V-range for voltage measurement and 2 mA or 20 mA for current measurement.

Next, from these data, draw two graphs, one is in linear scale (on a section paper) and the other is in semi-logarithmic style in which the vertical axis is the current and the horizontal axis is the voltage.

Table 8.6.

Red LED		Green LED		Blue LED	
Voltage V (V)	Current I (mA)	Voltage V (V)	Current I (mA)	Voltage V (V)	Current I (mA)
1.41 ± 0.01	0.010 ± 0.001	1.66 ± 0.01	0.015 ± 0.001	2.44 ± 0.01	0.010 ± 0.001
1.44	0.017	1.68	0.026	2.49	0.015
1.5	0.062	1.71	0.046	2.55	0.031
1.55	0.145	1.72	0.063	2.59	0.053
1.59	0.310	1.74	0.110	2.62	0.098
1.64	0.720	1.77	0.230	2.70	0.300
1.7	2.30 ± 0.01	1.80	0.460	2.77	0.820
1.73	5.80	1.83	1.05 ± 0.01	2.81	1.35 ± 0.01
1.8	10.4 ± 0.1	1.86	2.00	2.89	3.95
1.85	15.4	1.92	5.20	2.97	9.00
1.89	19.0	1.97	9.90	3.03	14.0 ± 0.1
		2.01	14.7 ± 0.1	3.11	19.9
		2.06	19.8		

Three data sets for these three LEDs should be plotted on a single graph. Each curve is called a "current-voltage characteristic curve" of the LED.

Cautions

1. The range of voltage should be determined by yourself. Be sure that the current should not exceed $20\,\text{mA}$. If it exceeds, the LED will be broken.
2. You should not measure the current larger than the maximum for each measurement range in the digital multi-meter. A fuse in it will be blown if the current exceeds the maximum. You should properly change the measurement range in the digital multi-meter.

Question 7 Here we define the threshold voltage V_0 used in Question 4 by the voltage that corresponds to a current of $0.1\,\text{mA}$, instead of by the voltage for lightening. Read the voltage V_0 of each LED from the semi-logarithmic graph, and calculate $E(= eV_0)$ for each LED. Next, draw a graph in which the vertical axis is E and the horizontal axis is ν. Then, by fitting the data with Eq. (8.6), obtain the value of Planck's constant h, together with the error.

Answer

Question 1 From the data in Table 8.1, the wavelength of each light is as follows:

$$\text{Red: } \lambda = 638\,\text{nm}; \quad \text{Green: } \lambda = 551\,\text{nm}; \quad \text{Blue: } \lambda = 493\,\text{nm}.$$

Question 2

From Eq. (8.5), we get $\lambda = \dfrac{xd}{\sqrt{L^2 + x^2}}$. Then by taking the derivative with x,

$$\frac{d\lambda}{dx} = \lambda \frac{L^2}{x(L^2 + x^2)}.$$

Due to the error Δx in the measurement of x, the error in λ is

$$\Delta\lambda = \left| \frac{d\lambda}{dx} \Delta x \right| = \left| \frac{\lambda L^2}{x(L^2 + x^2)} \Delta x \right|.$$

It is calculated for each LED:

Red: $\Delta\lambda = 7.7$ nm;   Green: $\Delta\lambda = 8.1$ nm;   Blue: $\Delta\lambda = 24.8$ nm.

Since the significant figure of x is two digits, the final results of the wavelength are

$$\text{Red: } \lambda \pm \Delta\lambda = 640 \pm 10 \text{ nm},$$

$$\text{Green: } \lambda \pm \Delta\lambda = 550 \pm 10 \text{ nm},$$

$$\text{Blue: } \lambda \pm \Delta\lambda = 490 \pm 20 \text{ nm}.$$

Question 3 Since Eq. (8.1) tells that $\nu = \frac{c}{\lambda}$, the error of frequency due to the error of wavelength is $\Delta\nu = |\frac{d\nu}{d\lambda}\Delta\lambda| = \frac{\nu}{\lambda}\Delta\lambda$. The frequency ν of each color of light is,

$$\text{Red: } 4.70 \times 10^{14} \text{ Hz};   \text{Green: } 5.44 \times 10^{14} \text{ Hz};   \text{and}$$

$$\text{Blue: } 6.09 \times 10^{14} \text{ Hz}.$$

The $\Delta\nu$ is calculated for each color;

Red: 0.06×10^{14} Hz;   Green: 0.08×10^{14} Hz;   Blue: 0.31×10^{14} Hz.

Then by considering the significant figure, the results are

$$\text{Red: } \nu \pm \Delta\nu = (4.7 \pm 0.1) \times 10^{14} \text{ Hz},$$

$$\text{Green: } \nu \pm \Delta\nu = (5.4 \pm 0.1) \times 10^{14} \text{ Hz},$$

$$\text{Blue: } \nu \pm \Delta\nu = (6.1 \pm 0.3) \times 10^{14} \text{ Hz}.$$

Question 4 Since $E \pm \Delta E = e(V_0 \pm \Delta V_0)$, the results are

Red: $E \pm \Delta E = 1.60 \times 10^{-19} \times (1.6 \pm 0.1) = (2.6 \pm 0.2) \times 10^{-19}$ J,

Green: $E \pm \Delta E = 1.60 \times 10^{-19} \times (1.8 \pm 0.1) = (2.9 \pm 0.2) \times 10^{-19}$ J,

Blue: $E \pm \Delta E = 1.60 \times 10^{-19} \times (2.6 \pm 0.1) = (4.2 \pm 0.2) \times 10^{-19}$ J.

Question 5

Although the data points are scattered, we fitted straight lines passing through the origin (Fig. 8.16). The gradient of the fitted line

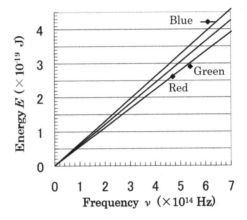

Fig. 8.16.

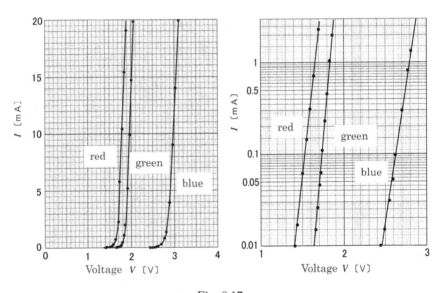

Fig. 8.17.

is Planck's constant h. From the central line, we get $h = \frac{4.25 \times 10^{-19}}{7.00 \times 10^{14}} =$ 6.07×10^{-34} J·sec. From the upper and lower lines, we estimated the error. Then, the final result is $h \pm \Delta h = (6.1 \pm 0.5) \times 10^{-34}$ J·sec.

Question 6 The data in Table 8.6 are plotted on linear scale and on semi-log scale, respectively, as shown in Fig. 8.17:

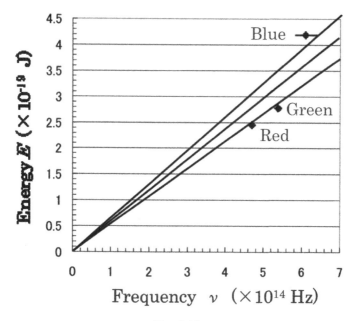

Fig. 8.18.

Question 7 The threshold voltage V_0 of each color LED is obtained from the semi-log graph in Question 6: Red: $V_0 = 1.53 \pm 0.01$ V; Green: $V_0 = 1.73 \pm 0.01$ V; Blue: $V_0 = 2.62 \pm 0.01$ V. Then the energy of light $E = eV_0$ is calculated: Red: $(2.45 \pm 0.02) \times 10^{-19}$ J; Green: $(2.77 \pm 0.02) \times 10^{-19}$ J, Blue: $(4.19 \pm 0.02) \times 10^{-19}$ J. Next, we drew a required graph as shown in Fig. 8.18. From the gradient of the fitted lines $E = h\nu$, we obtained the value of Planck's constant $h = (5.9 \pm 0.6) \times 10^{-34}$ J·sec. This is consistent with the result in Question 5.

Ref According to Chronological Scientific Tables, the value of Planck's constant is $h = 6.626 \times 10^{-34}$ J·sec.

Appendix

Mathematical Physics

Although the participants in the International Physics Olympiad (abbreviated IPhO) are not required to have the sophisticated skills for calculus, differential equations and complex number, these skills are extremely useful for describing and studying physics. Approximation formulae derived by the calculus may help us to extract physical essence from complicated phenomena. Furthermore, the employment of differential equation may enable us to find that apparently unrelated physical phenomena have analogous mathematical structures.

In this appendix, we outline the differential and integral calculus out of the mathematical physics which is not taught in the high-school education in Japan. It will help you understand well to solve problems with your own hands.

A.1. Inverse Trigonometric Functions

It is impossible to define the inverse trigonometric functions themselves because they are not single-valued functions. Hence, we restrict the domain of coordinates such that the values of functions and the coordinates have one-to-one correspondence. Then it is possible to define their inverse functions. For example, $\arcsin x$ and $\arctan x$ are the inverse functions of $\sin x$ and $\tan x$ in domain $[-\pi/2, \pi/2]$, respectively. The inverse function of $\cos x$ is denoted by $\arccos x$ in domain $[0, \pi]$.

The inverse trigonometric functions allow us to easily calculate some particular integrals which require the integration by substitution in the high-school mathematics in Japan.

Example A.1. Differentiate the following functions:

$$\arcsin x, \quad \arccos x, \quad \arctan x$$

Solution

Let $y = \arcsin x$, then from the definition we have $x = \sin y$. Thereby, differentiation of this equation yields

$$\frac{dy}{dx} = \frac{1}{dx/dy} = \frac{1}{\cos y} = \frac{1}{\sqrt{1 - x^2}} \quad \left(\because \cos y > 0, \ y \in \left[-\frac{\pi}{2}, \frac{\pi}{2} \right] \right).$$

Thus, we obtain

$$(\arcsin x)' = 1/\sqrt{1 - x^2}.$$

Similarly, we have

$$(\arccos x)' = -1/\sqrt{1 - x^2}, \quad (\arctan x)' = 1/(1 + x^2). \quad \blacksquare$$

From Example A.1, we obtain the following integral formulae.

$$\int \frac{dx}{\sqrt{a^2 - x^2}} = \arcsin \frac{x}{a} + \text{const.},$$

$$-\int \frac{dx}{\sqrt{a^2 - x^2}} = \arccos \frac{x}{a} + \text{const.}$$

$$\int \frac{dx}{a^2 + x^2} = \frac{1}{a} \arctan \frac{x}{a} + \text{const.}$$

Example A.2. Using the energy conservation law of harmonic oscillation $\frac{1}{2}mv^2 + \frac{1}{2}kx^2 = E$ (E is a constant), express dt/dx as a function in terms of x. Furthermore, integrate it to derive the relation between x and t.

Solution

From $v^2 = \frac{2E}{m} - \frac{k}{m}x^2$, we obtain $\frac{dx}{dt} = \sqrt{\frac{k}{m}} \sqrt{\frac{2E}{k} - x^2}$, or,

$$\frac{dt}{dx} = \sqrt{\frac{m}{k}} \frac{1}{\sqrt{\frac{2E}{k} - x^2}}.$$

Using the Integral formula derived in Ex. A.1, we have

$$t = \sqrt{\frac{m}{k}} \left(\arcsin \sqrt{\frac{k}{2E}} x - \varphi \right), \, (\varphi \text{ is an integral constant.}),$$

from which we obtain

$$x = \sqrt{\frac{2E}{k}} \sin \left(\sqrt{\frac{k}{m}} t + \varphi \right).$$ ■

A.2. Useful Coordinate Systems

The use of an appropriate coordinate system can simplify the calculation in physics problems. Here, we review three most useful coordinate system: The two-dimensional polar coordinate system, the cylindrical coordinate system and the spherical coordinate system (or the three-dimensional polar coordinate system).

A.2.1. *Two-Dimensional Polar Coordinate System*

In the two-dimensional polar coordinate system, a point P is expressed by (r, θ), where $r = \overrightarrow{OP}$, and θ is the angle between $\overrightarrow{OP}$ and the positive x-axis as shown in Fig. A.1. The relations between the two-dimensional polar coordinates, (r, θ), and the rectangular (Cartesian) coordinates, (x, y), are given by $x = r \cos \theta$ and $y = r \sin \theta$. The component along OP is called the r-component while the component

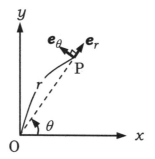

Fig. A.1.

in the counter clockwise direction perpendicular to $\overrightarrow{OP}$ is called the θ-component. The unit vectors in the polar coordinate system, e_r and e_θ, can be expressed as

$$e_r = \begin{pmatrix} \cos\theta \\ \sin\theta \end{pmatrix}, \quad e_\theta = \begin{pmatrix} -\sin\theta \\ \cos\theta \end{pmatrix}.$$

Example A.3. The position vector r is expressed in terms of (r, θ) and e_r as

$$r = \begin{pmatrix} x \\ y \end{pmatrix} = r\begin{pmatrix} \cos\theta \\ \sin\theta \end{pmatrix} = re_r.$$

Calculate the velocity vector $v = \dot{r}$ and the acceleration vector $a = \ddot{r}$ in the polar coordinate system. That is, find the polar components $v_r, v_\theta, a_r, a_\theta$ where

$$v = v_r e_r + v_\theta e_\theta, \quad a = a_r e_r + a_\theta e_\theta.$$

Solution

First, calculate the time derivative of the unit vectors:

$$\dot{e}_r = \begin{pmatrix} -\dot{\theta}\sin\theta \\ \dot{\theta}\cos\theta \end{pmatrix} = \dot{\theta}\,e_\theta, \quad \dot{e}_\theta = -\begin{pmatrix} \dot{\theta}\cos\theta \\ \dot{\theta}\sin\theta \end{pmatrix} = -\dot{\theta}\,e_r.$$

Then,

$$v = \dot{r} = \dot{r}e_r + r\dot{e}_r = \dot{r}e_r + r\dot{\theta}e_\theta$$

$$a = \dot{v} = \ddot{r}e_r + \dot{r}\dot{e}_r + \dot{r}\dot{\theta}e_\theta + r\ddot{\theta}e_\theta + r\dot{\theta}\dot{e}_\theta$$

$$= \ddot{r}e_r + \dot{r}\dot{\theta}e_\theta + \dot{r}\dot{\theta}e_\theta + r\ddot{\theta}e_\theta + r\dot{\theta}\cdot(-\dot{\theta}e_r)$$

$$= (\ddot{r} - r\dot{\theta}^2)e_r + (2\dot{r}\dot{\theta} + r\ddot{\theta})e_\theta,$$

from which we have $v_r = \underline{\dot{r}}$, $v_\theta = \underline{r\dot{\theta}}$, $a_r = \underline{\ddot{r} - r\dot{\theta}^2}$ and $a_\theta = \underline{2\dot{r}\dot{\theta} + r\ddot{\theta}}$.

∎

Example A.4. Suppose a planet, P, of mass m moves around the Sun, S, which is of a larger mass, M, due to the universal gravitation $F = G\frac{Mm}{r^2}$, as shown in Fig. A.2. Here, r and G denote the distance between the planet and the Sun and the gravitational

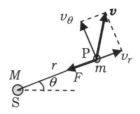

Fig. A.2.

constant, respectively. Since M is large, the Sun can be considered at rest.

(1) Show that the angular momentum $L = mr^2\dot\theta$ is conserved.
(2) Find the total energy E of the planet moving around the Sun in terms of r, its time derivative $\dot r$, G and L.

Find the condition for the total energy E that the planet continues turning around the Sun and does not move from the Sun to an infinite distance away.

Finally, supposing that the planet moves on a circular orbit around the Sun, find the relation between the radius of the planet orbit and the angular momentum L.

Solution

(1) The universal gravitation has no θ-direction component. Such a force is called the central force. The equation of motion in the θ-direction is

$$ma_\theta = 0 \quad \Rightarrow \quad 2\dot r\dot\theta + r\ddot\theta = 0.$$

Then,

$$\frac{dL}{dt} = \frac{d}{dt}(mr^2\dot\theta) = 2mr\dot r\dot\theta + mr^2\ddot\theta = mr(2\dot r\dot\theta + r\ddot\theta) = 0.$$

Namely, L is conserved in time.

(2) Suppose the mechanical energy of the planet moving around the Sun under the universal gravitation. Let the potential energy be zero at $r = \infty$. The potential energy U is obtained by the integration of the universal gravitation $-\frac{GMm}{r^2}$ from $r = r$ to

$r = \infty$:

$$U = \int_r^\infty \left(-\frac{GMm}{r^2} \right) dr = -\frac{GMm}{r}.$$

On the other hand, the kinetic energy K can be written in terms of the polar coordinates as

$$K = \frac{1}{2}mv^2 = \frac{1}{2}m \left(v_r^2 + v_\theta^2 \right)$$

$$= \frac{1}{2} \left\{ m\dot{r}^2 + m \left(r\dot{\theta} \right)^2 \right\}$$

$$= \frac{1}{2} \left(m\dot{r}^2 + \frac{L^2}{mr^2} \right).$$

Therefore, the total mechanical energy $E = K + U$ becomes

$$E = \frac{1}{2}m\dot{r}^2 + \frac{L^2}{2mr^2} - \frac{GMm}{r}. \tag{A.1}$$

The sum of the second term and the third term on the right hand side of Eq. (A.1) is called the **effective potential**:

$$U_0(r) = \frac{L^2}{2mr^2} - \frac{GMm}{r}.$$

Since the angular momentum of the planet, L, is constant, the right hand side of Eq. (A.1) is the sum of the kinetic energy in the r direction, $\frac{1}{2}m\dot{r}^2$, and the potential energy $U_0(r)$ which is determined by r. Hence Eq. (A.1) can be regarded as the mechanical energy of a one dimensional motion of a particle in the r direction.

Now, the derivative of the effective potential becomes

$$\frac{dU_0}{dr} = -\frac{L^2}{mr^3} + \frac{GMm}{r^2}.$$

The effective potential $U_0(r)$ has its negative minimum at $r = r_0 = \frac{L^2}{GMm^2}$. Note that when the distance between the planet and the Sun is infinity, $r \to \infty$, the effective potential goes to zero, $U_0(r) \to 0$. The outline of $U_0(r)$ is shown in Fig. A.3.

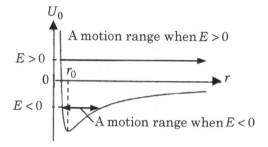

Fig. A.3.

Suppose the planet is moving away from the Sun ($\dot{r} > 0$) at some instant. If $E > 0, \dot{r}$ continues to be positive as r increases (see Fig. A.3) and the planet moves from the Sun to an infinite distance away.

On the other hand, if $E < 0$, as r increases, $\dot{r}$ of the planet becomes zero, i.e., $\frac{1}{2}m\dot{r}^2 = 0$. Thereafter, $\dot{r} < 0$ and the planet approaches the Sun. The planet goes back and forth between two points of $\dot{r} = 0$. Thereby the planet stays at a range of finite distance from the Sun. Therefore the condition that the planet continues turning around the Sun is given by $\underline{E < 0}$.

Since r is constant when the planet moves circularly, the radius should be $r = r_0$ in which effective potential $U_0(r)$ takes its minimum value. Hence, the relation between the radius r and the angular momentum L is

$$\underline{r = \frac{L^2}{GMm^2}}.$$ ■

A.2.2. *Cylindrical Coordinate System*

In the cylindrical coordinate system, a point $P(x, y, z)$ is expressed by (r, θ, z), where (r, θ) are the polar coordinates of the foot of perpendicular from P to the x–y plane, as shown in Fig. A.4. The volume element in this coordinate system is given as follows. Since the infinitesimal area dS in the region, $r \sim r + dr$ and $\theta \sim \theta + d\theta$, is expressed by $dS = r dr d\theta$ as shown in Fig. A.5, the infinitesimal volume dV in the region, $r \sim r + dr, \theta \sim \theta + d\theta$ and

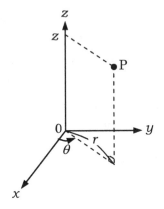

Fig. A.4.

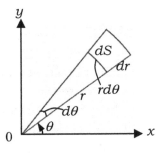

Fig. A.5.

$z \sim z + dz$, is expressed by $dV = rdrd\theta dz$. The volume integral of a function $f(r, \theta, z)$ expressed in terms of the cylindrical coordinates is written as

$$\iiint dV f(r, \theta, z) = \int dz \int d\theta \int dr [r \cdot f(r, \theta, z)],$$

where the right hand side in the above equation stands for the integral with respect to z, θ, r (see Example A.5) and is called a **multiple integral**.

Example A.5. Using the volume integration with respect to the cylindrical coordinates, find the moment of inertia I_1 of a uniform cylinder of radius R, height L and mass m around the central axis.

And also find the moment of inertia I_2 of a uniform sphere of radius R and mass m around an axis through the center.

Solution

Cylinder: Let the density be ρ. Then we have

$$m = \int_0^L dz \int_0^{2\pi} d\theta \int_0^R dr\,[r \cdot \rho] = L \cdot 2\pi \cdot \rho \cdot \frac{R^2}{2} = \pi R^2 L\rho,$$

$$I_1 = \int_0^L dz \int_0^{2\pi} d\theta \int_0^R dr\,[r \cdot \rho r^2] = L \cdot 2\pi \cdot \rho \cdot \frac{R^4}{4}$$

$$= \pi R^2 L\rho \cdot \frac{R^2}{2} = \frac{1}{2} m R^2.$$

Sphere: Let the density be ρ. For given $z(-R \le z \le R)$, the value of r is restricted to the region, $0 \le r \le \sqrt{R^2 - z^2}$. Hence we have

$$m = \int_{-R}^R dz \int_0^{2\pi} d\theta \int_0^{\sqrt{R^2-z^2}} dr\,[r \cdot \rho] = 2\pi\rho \int_{-R}^R \frac{R^2 - z^2}{2} dz$$

$$= \frac{4}{3}\pi R^3 \rho,$$

$$I_2 = \int_{-R}^R dz \int_0^{2\pi} d\theta \int_0^{\sqrt{R^2-z^2}} \cdot \, dr\,[r \cdot \rho r^2] = 2\pi\rho \int_{-R}^R \frac{(R^2 - z^2)^2}{2} dz$$

$$= \frac{8}{15}\pi R^5 \rho = \frac{2}{5} m R^2. \qquad \blacksquare$$

Compare the above calculation with that given in Example 2.13.

A.2.3. *Spherical Coordinate System*

In the spherical coordinate system, a point P (x, y, z) is expressed by (r, θ, ϕ), where r is the distance between the origin and the point P, θ is the angle between the positive z-axis and OP, and ϕ is the angle between the positive x-axis and the projection of OP to the x–y plane as shown in Fig. A.6.

Example A.6. Express (x, y, z) in terms of (r, θ, ϕ). Find the infinitesimal volume in the region, $r \sim r + dr, \theta \sim \theta + d\theta$ and $\phi \sim \phi + d\phi$.

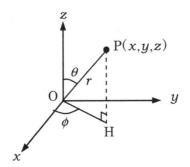

Fig. A.6.

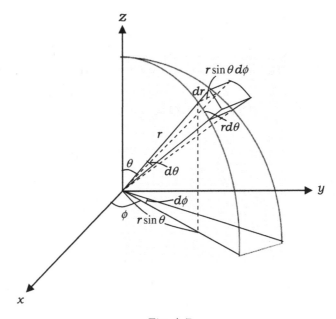

Fig. A.7.

Solution

From Fig. A.6, we have $x = r \sin \theta \cos \phi, y = r \sin \theta \sin \phi, z = r \cos \theta$. From Fig. A.7, the lengths of edges of the infinitesimal region are given by dr, $r d\theta$ and $r \sin \theta d\phi$, respectively. Therefore the infinitesimal volume can be expressed by

$$dV = r^2 \sin \theta dr d\theta d\phi.$$
∎

Example A.7. Suppose there is a point mass, m, at a distance r from the center of a uniform spherical shell of radius a and mass M. Find the expression for the gravitational potential energy of the point mass.

Solution

Let the center of the spherical shell be the origin. Then, the potential energy is spherically symmetric and depends only on r. Therefore, the potential energy of the point mass at the distance r from the origin is the same as that of the point mass at $P(0,0,r)$ in the Cartesian coordinates. As shown in Fig. A.8, the potential energy dU generated by an infinitesimal mass dM at a point $Q(a, \theta, \phi)$ is given by

$$dU = -\frac{GmdM}{r_1} = -\frac{GmdM}{\sqrt{r^2 - 2ar\cos\theta + a^2}},$$

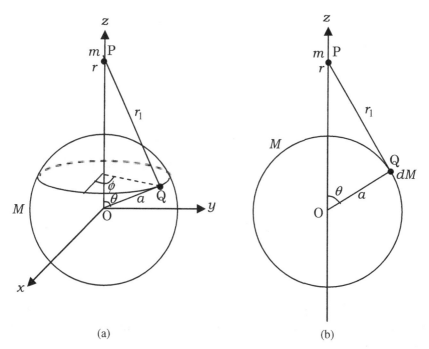

(a)   (b)

Fig. A.8.   (b) is a cross section by the plane including the z-axis and the point Q in (a).

where the length of the segment PQ, r_1, is expressed as $r_1 = \sqrt{r^2 - 2ar\cos\theta + a^2}$, by using the cosine theorem. The infinitesimal mass dM is given by

$$dM = \frac{M}{4\pi a^2} a^2 \sin\theta d\theta d\phi = \frac{M}{4\pi} \sin\theta d\theta d\phi.$$

Then, we can express the potential energy as follows:

$$U = \int_S dU = \int_0^{2\pi} d\phi \int_0^\pi d\theta \left[-\frac{GMm}{4\pi} \frac{\sin\theta}{\sqrt{r^2 - 2ra\cos\theta + a^2}} \right],$$

where S is the spherical surface.

Denoting $s = \cos\theta$, we get

$$U = 2\pi \cdot \frac{GMm}{4\pi} \int_1^{-1} \frac{ds}{\sqrt{r^2 - 2ras + a^2}}$$

$$= -\frac{GMm}{2} \int_{-1}^1 \frac{ds}{\sqrt{r^2 - 2ras + a^2}}.$$

Computation of the definite integral yields

$$\int_{-1}^1 \frac{ds}{\sqrt{r^2 - 2ras + a^2}} = \left[\frac{\sqrt{r^2 - 2ras + a^2}}{-ra} \right]_{s=-1}^{s=1}$$

$$= \frac{|r+a| - |r-a|}{ra} = \begin{cases} 2/r, r > a \\ 2/a, r < a. \end{cases}$$

We finally obtain

$$U = \begin{cases} -\dfrac{GMm}{r}, & r > a \\ -\dfrac{GMm}{a}, & r < a. \end{cases}$$

When $r < a$, $U(r)$ is constant, so that the gravitational force is zero in this region. The potential energy at a point outside the spherical shell is equivalent to that generated by a point mass M located at the origin. ∎

A.3. Taylor Expansion

Given a n-times differentiable function $f(x)$, the polynomial

$$P_n(x) = f(a) + f'(a)(x - a) + \frac{f''(a)}{2!}(x - a)^2$$

$$+ \cdots + \frac{f^{(n)}(a)}{n!}(x - a)^n = \sum_{k=0}^{n} \frac{f^{(k)}(a)}{k!}(x - a)^k,$$

is called the n-th order **Taylor polynomial** of $f(x)$ at $x = a$. When $(x - a)$ is sufficiently small, the difference between $P_n(x)$ and $f(x)$ is of the same order as $(x - a)^{n+1}$, so that we can approximate $f(x)$ by $P_n(x)$ to this order. Conversely, a polynomial different from $f(x)$ by the order $(x - a)^{n+1}$ is $P_n(x)$ alone. The function

$$T(x) = \lim_{n \to \infty} P_n(x) = \sum_{k=0}^{\infty} \frac{f^{(k)}(a)}{k!}(x - a)^k,$$

is called the **Taylor series** of $f(x)$ at $x = a$. When the Taylor series $T(x)$ converges and coincides with $f(x)$ everywhere in the domain of $f(x)$, it is said that $f(x)$ can be expanded in the **Taylor expansion**. When $a = 0$, the Taylor expansion is called the **Maclaurin expansion**.

Example A.8. Derive the Maclaurin expansion of $f(x) = \frac{1}{1-x}$ ($|x| < 1$).

Solution

Let us denote

$$f(x) = \frac{1}{1 - x} = a_0 + a_1 x + a_2 x^2 + a_3 x^3 + \cdots . \tag{A.2}$$

Substituting $x = 0$ into Eq. (A.2), we obtain $a_0 = 1$. Differentiating Eq. (A.2) with respect to x, we have

$$f'(x) = \frac{1}{(1 - x)^2} = a_1 + 2a_2 x + 3a_3 x^2 + \cdots + k a_k x^{k-1} + \cdots . \tag{A.3}$$

Substituting $x = 0$ into Eq. (A.3), we obtain $a_1 = 1$.

In the same way, "Differentiate and substitute $x = 0$ into the equation derived". Differentiating $f(x)$ k times, we get

$$f^{(k)}(x) = \frac{k!}{(1-x)^{k+1}} = k!a_k + (k+1)!a_{k+1}x + \cdots,$$

and substituting $x = 0$ into the above equation, we can get the coefficient of x^k as $a_k = 1$. Hence, the Maclaurin expansion of $f(x)$ becomes

$$f(x) = \frac{1}{1-x} = 1 + x + x^2 + \cdots = \sum_{k=0}^{\infty} x^k. \qquad \text{(A.4)}$$

∎

Another Solution

The identity,

$$(1-x)(1+x+\cdots+x^k) = 1 - x^{k+1},$$

derives the following

$$1 + x + \cdots + x^k = \frac{1 - x^{k+1}}{1-x}.$$

The difference between $f(x) = \frac{1}{1-x}$ and $1+x+\cdots+x^k$ is of the order x^{k+1} when $|x| < 1$. Thus, $1 + x + \cdots + x^k$ is the Taylor polynomial, and taking the limit $k \to \infty$ we have the Maclaurin expansion, Eq. (A.4). ∎

Taylor expansions of rational functions can often be derived easily as above.

Example A.9. Derive the Maclaurin expansions for $\sin x$, $\cos x$, and e^x.

Solution

Differentiating $f(x) = \sin x$ with respect to x, we have

$$f'(x) = \cos x, \quad f''(x) = -\sin x, \quad f'''(x) = -\cos x,$$
$$f^{(4)}(x) = \sin x, \ldots$$

In general, $f^{(2k)}(x) = (-1)^k \sin x$, $f^{(2k+1)}(x) = (-1)^k \cos x$. Thus, $f^{(2k)}(0) = 0$, $f^{(2k+1)}(0) = (-1)^k$.

Suppose $f(x) = a_0 + a_1 x + a_2 x^2 + \cdots = \sum_{k=0}^{\infty} a_k x^k$, then we have $a_0 = 0, a_1 = 1, a_2 = 0, a_3 = -\frac{1}{3!}, \ldots$, and obtain

$$\sin x = x - \frac{1}{3!} x^3 + \cdots = \sum_{k=0}^{\infty} \frac{(-1)^k}{(2k+1)!} x^{2k+1}.$$

In the same way, the Maclaurin expansion of $\cos x$ is obtained as

$$\cos x = 1 - \frac{1}{2!} x^2 + \cdots = \sum_{k=0}^{\infty} \frac{(-1)^k}{(2k)!} x^{2k}.$$

Note that since $\sin x$ is an odd function of x, it is expanded in a series of odd power of x, x^{2k+1}. In the same way, since $\cos x$ is an even function of x, it is expanded in a series of even power of x, x^{2k}.

Next, we derive the Maclaurin expansion of $f(x) = e^x$. Since

$$f'(x) = f''(x) = \cdots = f^{(k)}(x) = \cdots = e^x$$
$$\Rightarrow f'(0) = f''(0) = \cdots = f^{(k)}(0) = \cdots = 1,$$

the Maclaurin expansion of e^x is given by

$$e^x = 1 + x + \frac{1}{2!} x^2 + \cdots = \sum_{k=0}^{\infty} \frac{1}{k!} x^k. \qquad \blacksquare$$

A.4. Taylor Polynomials as Approximation Formulae

Taylor polynomials provide a given function with approximate expression at particular accuracy. Therefore, they are useful in order to calculate physical quantities with the required accuracy.

Example A.10. Derive the second order approximation formula (Taylor polynomial) of $f(x) = (1 + x)^a$.

Solution

$$f'(x) = a(1+x)^{a-1}, \quad f''(x) = a(a-1)(1+x)^{a-2},$$
$$\Rightarrow f'(0) = a, \quad f''(0) = a(a-1).$$

The second order Taylor expansion is

$$(1 + x)^a \approx 1 + ax + \frac{a(a-1)}{2}x^2. \qquad \blacksquare$$

Example A.11. Suppose there are a couple of electric charges q and $-q$, respectively, at $(0, d)$ and $(0, -d)$ in vacuum, as shown in Fig. A.9. Derive the potential V at (x, y), which is located sufficiently far from the origin, up to the first order with respect to d. Let the potential at infinity be zero, and denote the vacuum permittivity by ε_0.

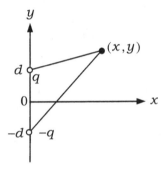

Fig. A.9.

Solution

By definition,

$$V = \frac{q}{4\pi\varepsilon_0} \frac{1}{\sqrt{x^2 + (y-d)^2}} - \frac{q}{4\pi\varepsilon_0} \frac{1}{\sqrt{x^2 + (y+d)^2}}.$$

On the other hand, using the approximation formula in Example A.10, we have

$$\frac{1}{\sqrt{x^2 + (y-d)^2}} = \frac{1}{\sqrt{x^2 + y^2}} \left(1 - \frac{2yd - d^2}{x^2 + y^2} \right)^{-\frac{1}{2}}$$

$$\approx \frac{1}{\sqrt{x^2 + y^2}} \left(1 + \frac{yd - d^2/2}{x^2 + y^2} \right)$$

$$\approx \frac{1}{\sqrt{x^2 + y^2}} \left(1 + \frac{yd}{x^2 + y^2} \right).$$

In the same way,

$$\frac{1}{\sqrt{x^2 + (y+d)^2}} = \frac{1}{\sqrt{x^2 + y^2}}\left(1 - \frac{yd}{x^2 + y^2}\right).$$

Therefore, we obtain

$$V = \frac{q}{2\pi\varepsilon_0}\frac{yd}{\left(x^2 + y^2\right)^{\frac{3}{2}}}. \qquad \blacksquare$$

Example A.12. In the presence of a uniform and weak magnetic field B, a particle of charge q and mass m with velocity v is shot toward a point P on the screen which is located at distance L apart from the initial position of the particle, as shown in Fig. A.10. The incident particle is moving perpendicularly to the screen. Derive the distance l between P and the point where the particle actually hits the screen. Assume that the deflection angle of the particle (which indicates the change in the advancing direction) is sufficiently small. You may use approximation formulae: $\sin\theta \approx \tan\theta \approx \theta$, and $\cos\theta \approx 1 - \theta^2/2$ ($|\theta| \ll 1$), if necessary.

Solution

Because the Lorentz force acts perpendicularly to the velocity of the particle, it does not do any work on the particle. Hence the speed of circular motion of the particle remains in constant value v. The centripetal force of the uniform circular motion is supplied by the Lorentz force. So, let r be the radius of the circular motion, then we

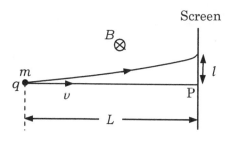

Fig. A.10.

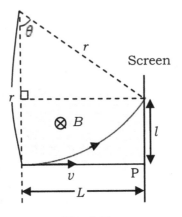

Fig. A.11.

have

$$m\frac{v^2}{r} = qvB \quad \therefore \ r = \frac{mv}{qB}.$$

Suppose the direction of the particle is changed by angle θ before it reaches the screen as shown in Fig. A.11, then we have

$$l = r - r\cos\theta \approx \frac{r\theta^2}{2}.$$

On the other hand, $\sin\theta = L/r$, then $\theta \approx L/r$, hence,

$$l \approx \frac{L^2}{2r} = \frac{qBL^2}{2mv}.$$ ■

A.5. Complex Plane

As shown in Fig. A.12, the two-dimensional Cartesian coordinates, (x, y), have one-to-one correspondence to the complex number $z = x + iy$, where i is the imaginary unit ($i^2 = -1$). Therefore the x–y plane can be considered identical with the whole complex numbers. A complex number z is expressed in terms of the polar coordinates (r, θ) as

$$z = r(\cos\theta + i\sin\theta).$$

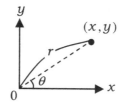

Fig. A.12.

Example A.13. Let $z_1 = \cos\theta_1 + i\sin\theta_1$ and $z_2 = \cos\theta_2 + i\sin\theta_2$. Calculate the product of z_1 and z_2, and express it in terms of the polar coordinates.

Solution

The additional theorem of trigonometric functions is written as

$$\cos(\theta + \phi) = \cos\theta\cos\phi - \sin\theta\sin\phi,$$
$$\sin(\theta + \phi) = \sin\theta\cos\phi + \cos\theta\sin\phi.$$

Using these formulae, we have

$$z_1 z_2 = \cos\theta_1\cos\theta_2 - \sin\theta_1\sin\theta_2 + i\sin\theta_1\cos\theta_2 + i\sin\theta_2\cos\theta_1$$
$$= \cos(\theta_1 + \theta_2) + i\sin(\theta_1 + \theta_2). \qquad\blacksquare$$

Example A.13 shows that the function of θ, $z = \cos\theta + i\sin\theta$, obeys the law of exponent (that is, the product of functions is given by the sum of variables included in the functions). In the next section, we will see that the exponential function and the trigonometric functions are connected with each other by the relation

$$e^{i\theta} = \cos\theta + i\sin\theta.$$

A.6. Euler's Formula

We naturally extend the domain of polynomial functions to complex numbers, by replacing real variables with complex variables. In the same way, we can extend the domain of expandable functions in Taylor series to complex numbers. For example, the trigonometric functions and the exponential function of complex variables can naturally be defined as

$$\sin z = \sum_{k=0}^{\infty} \frac{(-1)^k}{(2k+1)!} z^{2k+1}, \quad \cos z = \sum_{k=0}^{\infty} \frac{(-1)^k}{(2k)!} z^{2k},$$

$$e^z = \sum_{k=0}^{\infty} \frac{1}{k!} z^k. \tag{A.5}$$

(See Example A.9).

Differentiation of polynomials can also be extended naturally to complex variables. Complex derivatives of functions can be defined by differentiating the Taylor series of their functions term by term.

Example A.14. Confirm the relations $(\sin z)' = \cos z, (\cos z)' = -\sin z, (e^z)' = e^z$ by using Eq. (A.5).

Solution

$$(\sin z)' = \sum_{k=0}^{\infty} \left(\frac{(-1)^k}{(2k+1)!} z^{2k+1} \right)' = \sum_{k=0}^{\infty} \frac{(-1)^k}{(2k)!} z^{2k} = \cos z,$$

$$(\cos z)' = \left(\sum_{k=0}^{\infty} \frac{(-1)^k}{(2k)!} z^{2k} \right)' = \sum_{k=1}^{\infty} \frac{(-1)^k}{(2k-1)!} z^{2k-1}$$

$$= -\sum_{k'=0}^{\infty} \frac{(-1)^{k'}}{(2k'+1)!} z^{2k'+1} = -\sin z,$$

$$(e^z)' = \left(\sum_{k=0}^{\infty} \frac{z^k}{k!} \right)' = \sum_{k=1}^{\infty} \frac{z^{k-1}}{(k-1)!} = \sum_{k'=0}^{\infty} \frac{z^{k'}}{k'!} = e^z. \qquad \blacksquare$$

Example A.15. Confirm Euler's formula "$e^{i\theta} = \cos\theta + i\sin\theta$". Show that the trigonometric additional theorem can be derived from the law of exponents, and that the derivatives of the trigonometric functions can be derived from the derivative of the exponential function.

Solution

Let θ be a real number.

- Euler's formula:

$$\cos\theta + i\sin\theta = \sum_{k=0}^{\infty} \frac{(-1)^k}{(2k)!}\theta^{2k} + i\sum_{k=0}^{\infty} \frac{(-1)^k}{(2k+1)!}\theta^{2k+1}$$

$$= \sum_{k=0}^{\infty} \frac{i^{2k}}{(2k)!}\theta^{2k} + \sum_{k=0}^{\infty} \frac{i^{2k+1}}{(2k+1)!}\theta^{2k+1}$$

$$= \sum_{k=0}^{\infty} \frac{(i\theta)^k}{k!} = e^{i\theta}.$$

- The additional theorems:
  The relation $e^{i(\theta+\phi)} = e^{i\theta} \cdot e^{i\phi}$ leads to

$$\cos(\theta+\phi) + i\sin(\theta+\phi) = (\cos\theta + i\sin\theta)(\cos\phi + i\sin\phi).$$

Expanding the right-hand side and equating the real and imaginary parts of both sides of the equation separately, we obtain

$$\cos(\theta+\phi) = \cos\theta\cos\phi - \sin\theta\sin\phi,$$

$$\sin(\theta+\phi) = \sin\theta\cos\phi + \cos\theta\sin\phi.$$

- The derivatives:

$$(\cos\theta + i\sin\theta)' = (e^{i\theta})' = ie^{i\theta} = i(\cos\theta + i\sin\theta) = i\cos\theta - \sin\theta.$$

Comparing the real and imaginary parts separately, we obtain

$$(\cos\theta)' = -\sin\theta, \quad (\sin\theta)' = \cos\theta. \qquad \blacksquare$$

In physical calculations, expressing the trigonometric functions in terms of the complex exponential functions is often very useful.

Example A.16. There is a diffraction grating with N slits and a grating constant (a distance between adjacent slits) d. Applying a monochromatic light of wave length λ to this grating, we study the resulting diffraction pattern (Fig. A.13). Let ϕ be the phase difference between two adjacent slits. When the oscillation of the light wave from the first slit is expressed by $A\cos\omega t$, then that from the n-th slit can be expressed by $A\cos(\omega t - (n-1)\phi)$. The superposition of the oscillation from all the slits represents the actual oscillation. How

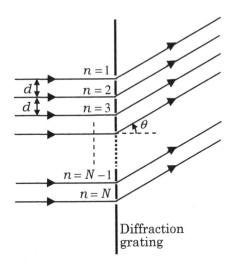

Fig. A.13.

does the intensity of the diffraction light depend on the diffraction angle θ.

Solution

Use the relation,

$$A\cos\omega t = \mathrm{Re}(Ae^{i\omega t}) = \frac{A}{2}(e^{i\omega t} + e^{-i\omega t}),$$

($\mathrm{Re}(z)$ represents the real part of z). Hereafter, we denote e^x as $\exp(x)$. Thus we have

$$\sum_{n=1}^{N} A\exp(i\omega t - i(n-1)\phi)$$

$$= Ae^{i\omega t}\sum_{n=1}^{N}\exp(-i(n-1)\phi) = Ae^{i\omega t}\frac{1-e^{-iN\phi}}{1-e^{-i\phi}},$$

$$\sum_{n=1}^{N} A\cos(\omega t - (n-1)\phi)$$

$$= \mathrm{Re}\left(\sum_{n=1}^{N} A\exp(i\omega t - i(n-1)\phi)\right)$$

$$= \frac{A}{2} \left(e^{i\omega t} \frac{1 - e^{-iN\phi}}{1 - e^{-i\phi}} + e^{-i\omega t} \frac{1 - e^{iN\phi}}{1 - e^{i\phi}} \right)$$

$$= \frac{A}{2} \frac{e^{\frac{iN\phi}{2}} - e^{-\frac{iN\phi}{2}}}{e^{\frac{i\phi}{2}} - e^{-\frac{i\phi}{2}}} \left(e^{i\omega t - \frac{iN\phi}{2} + \frac{i\phi}{2}} + e^{-i\omega t + \frac{iN\phi}{2} - \frac{i\phi}{2}} \right)$$

$$= A \frac{\sin N\phi/2}{\sin \phi/2} \cos \left(\omega t - \frac{N-1}{2} \phi \right).$$

Here, the amplitude of the composite wave is described by $A \frac{\sin N\phi/2}{\sin \phi/2}$, while its oscillation is described by $\cos(\omega t - \frac{N-1}{2}\phi)$. Because the intensity of wave is proportional to the square of the wave amplitude, the intensity is proportional to $(\frac{\sin N\phi/2}{\sin \phi/2})^2$. Since the light waves from two adjacent slits have the phase difference $\phi = 2\pi d \sin \theta / \lambda$, the intensity is proportional to,

$$\left(\frac{\sin(N\pi d \sin \theta / \lambda)}{\sin(\pi d \sin \theta / \lambda)} \right)^2. \qquad \blacksquare$$

A.7. Differential Equations 1 (Separation of Variables)

A differential equation is a mathematical equation which involves derivatives of unknown function as well as the function itself. The function which satisfies a differential equation is called its **solution**. The following type of differential equation

$$\frac{dy}{dx} = P(x)Q(y),$$

can be solved by separation of variables. This type of equation is called the **differential equation with separable variables**.

Dividing both sides of the above equation by the function $Q(y)$ and integrating both sides over x, we have

$$\int \frac{1}{Q(y)} \frac{dy}{dx} dx = \int P(x) dx \quad \Rightarrow \quad \int \frac{dy}{Q(y)} = \int P(x) dx.$$

The problem is then reduced to two indefinite integrals.

Example A.17. Solve the following differential equation:

$$\frac{dy}{dx} + P(x)y = 0.$$

Solution

Integrating both sides of the equation $\frac{1}{y}\frac{dy}{dx} = -P(x)$ over x, we have

$$\log|y| = -\int P(x)dx + C \, (C\text{: an integral constant}).$$

Denoting $\pm e^C$ by C_0, we have

$$y = C_0 \exp\left(-\int P(x)dx\right). \qquad (A.6)$$

In particular, $y \equiv 0$ (when $C_0 = 0$) is a solution of the differential equation. ∎

A.8. Differential Equations 2 (Linear)

Differential equations which contain a dependent variable and its derivatives up to the first-order with respect to all of them are called **linear differential equations**. In other words, a linear differential equation is expressed as

$$P_n(x)\frac{d^n y}{dx^n} + P_{n-1}(x)\frac{d^{n-1}y}{dx^{n-1}} + \cdots + P_1(x)\frac{dy}{dx} + P_0(x)y = Q(x),$$

$$(P_n(x) \not\equiv 0),$$

where n is called the **order** of the differential equation. When $Q(x) \equiv 0$, the equation is **homogeneous**; when $Q(x) \not\equiv 0$ it is **inhomogeneous**. Dividing the equation by $P_n(x)$ reduces the coefficient of the n-th order derivative to unity, and therefore we can hereafter assume $P_n(x) \equiv 1$. It is known that general solutions of n-th order differential equations include n integral constants.

A homogeneous first-order linear differential equation can be solved by separation of variables (Example A.17).

Example A.18. Let us consider an inhomogeneous first-order liner differential equation:

$$\frac{dy}{dx} + P(x)y = Q(x). \qquad (A.7)$$

Replacing the constant C_0 in Eq. (A.6) by a function, $C(x)$, we have

$$y = C(x) \exp\left(-\int P(x)dx\right)$$

Find $C(x)$ in the above.

Solution

Substituting

$$\frac{dy}{dx} = C'(x) \exp\left(-\int P(x)dx\right) - P(x)C(x) \exp\left(-\int P(x)dx\right)$$

into Eq. (A.7), we obtain

$$C'(x) \exp\left(-\int P(x)dx\right) = Q(x),$$

$$\therefore \quad C'(x) = Q(x) \exp\left(\int P(x)dx\right),$$

which can be integrated as

$$C(x) = \int Q(x) \exp\left(\int P(x)dx\right)dx + C_1$$

$$\overline{\qquad\qquad\qquad\qquad\qquad\qquad\qquad}$$

(C_1: an integral constant). ∎

Because $C(x)$ has one integral constant, a function,

$$y = \left(\int Q(x) \exp\left(\int P(x)dx\right)dx + C_1\right) \exp\left(-\int P(x)dx\right),$$

$$\text{(A.8)}$$

is the general solution of the inhomogeneous first- order liner differential equation.

In order to find solutions of the inhomogeneous differential equation, we replace the integral constants in the solution of corresponding homogeneous differential equation by functions of variables. This is called the **method of constant variation**.

We next consider second-order linear differential equations with constant coefficients.

Example A.19. For a second order differential equation,

$$\frac{d^2y}{dx^2} + 2p\frac{dy}{dx} + qy = 0, \tag{A.9}$$

find a solution in the form $y = e^{\alpha x}$.

Solution

Substituting $y = e^{\alpha x}$ into Eq. (A.9) and dividing its result by $e^{\alpha x}$, we obtain

$$\alpha^2 + 2p\alpha + q = 0. \tag{A.10}$$

Solutions of this quadratic equation are

$$\alpha_\pm = -p \pm \sqrt{p^2 - q}.$$

Then we obtain two solutions for Eq. (A.9) as follows:

$$y = \exp\left((-p \pm \sqrt{p^2 - q})x\right)$$

(Note that when $p^2 = q$, the two solutions are degenerate.) ∎

When $y = y_1, y_2$ are solutions of the homogeneous linear differential equation (A.9), a linear combination of y_1 and y_2,

$$y = C_1y_1 + C_2y_2,$$

is also a solution of Eq. (A.9). Because a solution with two integral constants is the general solution of the second-order differential equation, $y = C_+e^{\alpha_+x} + C_-e^{\alpha_-x}$ is the general solution of the differential equation (A.9) when $p^2 \neq q$. When $p^2 < q$, $\alpha_\pm$ is imaginary; thus using Euler's formula, we can rewrite the general solution as

$$y = e^{-px}\left(A\cos\sqrt{q - p^2}x + B\sin\sqrt{q - p^2}x\right),$$

where A and B are two arbitrary constants. This function represents an oscillation with the amplitude increasing or decreasing exponentially.

Example A.20. When $p^2 = q$, derive the general solution of Eq. (A.9) by using the method of constant variation.

Solution

Substituting $y = C(x)e^{-px}$ into the differential equation (A.9) and dividing its result by e^{-px}, we have

$$(C''(x) - 2pC'(x) + p^2 C(x)) + 2p(C'(x) - pC(x)) + p^2 C(x) = 0,$$

$$\therefore \ C''(x) = 0.$$

Thus we obtain

$$C(x) = \alpha x + \beta \quad (\alpha, \beta \text{ are integral constants}).$$

Then, the general solution is written as

$$y = (\alpha x + \beta)e^{-px}. \qquad \blacksquare$$

Consider an inhomogeneous second-order differential equation:

$$\frac{d^2 y}{dx^2} + 2p\frac{dy}{dx} + qy = f(x). \tag{A.11}$$

Simple calculation shows that the sum of a solution of this differential equation and a solution of the corresponding homogeneous differential equation (A.9) (i.e., the case of $f(x) \equiv 0$) is also a solution of the inhomogeneous differential equation (A.11). Therefore, if we find one of the solutions of Eq. (A.11), which is called a **particular solution**, we can obtain its general solution as the sum of the particular solution of Eq. (A.11) and the general solution of Eq. (A.9). When $f(x)$ is of a simple form, a particular solution can be obtained intuitively.

Example A.21. Let $f(x) = kx + l$, $q \neq 0$ and find a particular solution of Eq. (A.11).

Solution

Let $y = \alpha x + \beta$ be a particular solution of the differential equation (A.11). Then by substituting $y = \alpha x + \beta$ into Eq. (A.11), we have

$$q\alpha x + 2p\alpha + q\beta = kx + l,$$

which must be an identity, so that we obtain the relation

$$\alpha = \frac{k}{q}, \quad \beta = \frac{l}{q} - \frac{2kp}{q^2}.$$

Therefore, we find that

$$y = \frac{k}{q}x + \frac{l}{q} - \frac{2kp}{q^2},$$

is a particular solution of Eq. (A.11). ∎

Example A.22. Find a particular solution of Eq. (A.11) by using the method of constant variation.

Solution

Substituting $y = C(x)e^{\alpha x}(\alpha = \alpha_\pm)$ into the differential equation (A.11) and dividing its result by $e^{\alpha x}$, we obtain

$$(C''(x) + 2\alpha C'(x) + \alpha^2 C(x)) + 2p(C'(x) + \alpha C(x)) + qC(x)$$
$$= f(x)\,e^{-\alpha x}.$$

From Eq. (A.10), we have

$$C''(x) + 2(p + \alpha)C'(x) = f(x)\,e^{-\alpha x}.$$

This equation is a first-order liner differential equation for $C'(x)$. Thus one of the solutions can be written as

$$C'(x) = e^{-2(p+\alpha)x} \int f(x)e^{(2p+\alpha)x}dx.$$

By integration we get

$$C(x) = \int e^{-2(p+\alpha)x} \left(\int f(x)e^{(2p+\alpha)x}dx \right) dx.$$

Therefore, we find that

$$y = e^{\alpha x} \int e^{-2(p+\alpha)x} \left(\int f(x)e^{(2p+\alpha)x}dx \right) dx,$$

is a particular solution. ∎

A.9. Partial Differential Equation

Let $f(x_1, x_2, \ldots x_n)$ be a function of n independent variables $x_1, x_2, \ldots x_n$. It is called the **partial differentiation** to differentiate f with respect to x_i while keeping the other variables fixed. The derivative given by this procedure is written as $\frac{\partial f}{\partial x_i}$. Higher order partial derivatives are defined in the same way.

Equations for $f(x_1, x_2, \ldots x_n)$ and its partial derivatives are called **partial differential equations**. On the other hand, differential equations with a single variable, which are described in Secs. A.7 and A.8, are called **ordinary differential equations**. It is known that the general solutions of n-th order partial differential equations contain n arbitrary functions.

Example A.23. A function of two variables of real numbers, $f(x, y)$, satisfies a partial differential equation:

$$\frac{\partial^2 f}{\partial x \partial y} = 0.$$

Then, find $f(x, y)$.

$$\text{Note: } \frac{\partial^2 f}{\partial x \partial y} = \frac{\partial}{\partial x}\left(\frac{\partial f}{\partial y}\right).$$

Solution

$\frac{\partial}{\partial x}\left(\frac{\partial f}{\partial y}\right) = 0$ implies that $\frac{\partial f}{\partial y}$ is independent of x. Therefore, we have

$$\frac{\partial f}{\partial y} = \phi(y)(\phi(y) \text{ is an arbitrary function.}).$$

Then,

$$f = \psi(x) + \int \phi(y)dy = \underline{\psi(x) + \Phi(y)}. \tag{A.12}$$

Here, $\psi(x)$ and $\Phi(y) = \int \phi(y)dy$ are arbitrary functions of x and y, respectively and $\psi(x)$ comes out as an integral constant when f is considered a function of y. ∎

Although it is difficult to solve partial differential equations generally, some kind of partial differential equations can be solved by separating variables or by reducing the result to ordinary differential equations.

Example A.24. A function of two variables of real numbers, $f(x, y)$, satisfies a partial differential equation:

$$\frac{\partial f}{\partial x} + \frac{\partial f}{\partial y} = f. \tag{A.13}$$

Assuming $f(x, y) = X(x)Y(y)$, find the solution of Eq. (A.13).

Solution

Substituting $f(x, y) = X(x)Y(y)$ into Eq. (A.13), we have

$$Y\frac{\partial X}{\partial x} + X\frac{\partial Y}{\partial y} = XY,$$

which yields a relation

$$\frac{1}{X}\frac{\partial X}{\partial x} = 1 - \frac{1}{Y}\frac{\partial Y}{\partial y}.$$

The left-hand side is a function of x alone, and the right-hand side is a function of y alone. In order to make this equation valid in the entire domain of real numbers, the values of both sides must be constant and equal to each other. This fact leads to two ordinary differential equations:

$$\frac{1}{X}\frac{\partial X}{\partial x} = c_x, \quad \frac{1}{Y}\frac{\partial Y}{\partial y} = c_y, (c_x + c_y = 1),$$

$$\therefore \quad \frac{\partial X}{\partial x} = c_x X, \quad \frac{\partial Y}{\partial y} = c_y Y,$$

from which we can obtain

$$X = C_x \exp(c_x x), \quad Y = C_y \exp(c_y y).$$

Here, C_x and C_y are integral constants.

Thus, the solution can be written as

$$f = C \exp(c_x x + c_y y), \quad (c_x + c_y = 1, C = C_x C_y). \qquad \blacksquare$$

Let us consider the following partial differential equation.

$$\frac{\partial^2 f}{\partial t^2} - c^2 \frac{\partial^2 f}{\partial x^2} = 0 \quad (c > 0 \text{ is a constant}). \tag{A.14}$$

This is called the **one-dimensional wave equation**, which represents many physical phenomena such as oscillations of strings.

You can check that the function can describe the superposition of two waves moving toward each other,

$$f(x,\ t) = A \sin 2\pi \left(\frac{x}{\lambda} + ft\right) + B \sin 2\pi \left(\frac{x}{\lambda} - ft\right)$$

$$= A \sin \frac{2\pi}{\lambda}(x + ct) + B \sin \frac{2\pi}{\lambda}(x - ct),\ (c = f\lambda), \tag{A.15}$$

satisfies Eq. (A.14). Here, A and B are constant, λ is the wavelength and f denotes the frequency, respectively.

Example A.25. Transform the variables x and y to $\xi = x + ct$ and $\eta = x - ct$, and find the general solution of the wave equation (A.14). You can use the following formulas for the partial differentiation:

$$\frac{\partial f}{\partial x} = \frac{\partial f}{\partial \xi}\frac{\partial \xi}{\partial x} + \frac{\partial f}{\partial \eta}\frac{\partial \eta}{\partial x}, \quad \frac{\partial f}{\partial t} = \frac{\partial f}{\partial \xi}\frac{\partial \xi}{\partial t} + \frac{\partial f}{\partial \eta}\frac{\partial \eta}{\partial t} \tag{A.16}$$

Solution

Using Eq. (A.16), we have

$$\frac{\partial f}{\partial t} = \frac{\partial f}{\partial \xi}\frac{\partial \xi}{\partial t} + \frac{\partial f}{\partial \eta}\frac{\partial \eta}{\partial t} = c\left(\frac{\partial f}{\partial \xi} - \frac{\partial f}{\partial \eta}\right),$$

$$\frac{\partial^2 f}{\partial t^2} = c\left[\frac{\partial}{\partial \xi}\left(\frac{\partial f}{\partial t}\right) - \frac{\partial}{\partial \eta}\left(\frac{\partial f}{\partial t}\right)\right]$$

$$= c^2\left(\frac{\partial^2 f}{\partial \xi^2} - 2\frac{\partial^2 f}{\partial \xi \partial \eta} + \frac{\partial^2 f}{\partial \eta^2}\right).$$

Similarly we have

$$\frac{\partial^2 f}{\partial x^2} = \frac{\partial^2 f}{\partial \xi^2} + 2\frac{\partial^2 f}{\partial \xi \partial \eta} + \frac{\partial^2 f}{\partial \eta^2}.$$

Substituting these relations into the wave equation (A.14), we obtain

$$\frac{\partial^2 f}{\partial \xi \partial \eta} = 0$$

Therefore from Eq. (A.12), we obtain the solution in the form

$$f = \psi(\xi) + \phi(\eta),$$

where $\psi(\xi)$ and $\phi(\eta)$ are arbitrary functions of ξ and η, respectively, or explicitly,

$$f = \psi(x + ct) + \phi(x - ct). \tag{A.17}$$

■

Equation (A.17) corresponds to the superposition of sine waves as given in Eq. (A.15). Here $\xi = x + ct$ is proportional to the **phase** of $\psi(\xi)$, and is constant at the **equiphase points**. Differentiating ξ with respect to time t at the equiphase points, we obtain

$$\frac{dx}{dt} = -c.$$

This means that the equiphase points of $\psi(\xi)$ propagate with the speed c in the $-x$ direction. In the same way, the equiphase points of $\phi(\eta)$ propagate with the speed c in the $+x$ direction. The solution of the wave equation is a superposition of waves propagating along the x-axis with the speed c in the positive and negative directions.

A.10. Differential Equations and Physics

We consider some specific physical problems using the knowledge of differential equations described above.

Example A.26. A particle of mass m is moving with initial velocity v_0 along the x-axis. A resistance force of magnitude $\alpha v + \beta v^2$ acts on a particle with velocity v in the direction opposite to the velocity. Here α and β are constants. Write v as a function of t.

Solution

The equation of motion for the particle is

$$m\frac{dv}{dt} = -\alpha v - \beta v^2, \quad \therefore \quad \frac{1}{\alpha v + \beta v^2}\frac{dv}{dt} = -\frac{1}{m}.$$

Integrating both sides over t, we obtain

$$\int \frac{dv}{v(\alpha + \beta v)} = -\frac{1}{m}t + C.$$

Here, the left-hand side is calculated as

$$\int \frac{dv}{v(\alpha + \beta v)} = \int \frac{1}{\alpha}\left(\frac{1}{v} - \frac{\beta}{\alpha + \beta v}\right)dv$$

$$= \frac{1}{\alpha}\left(\log|v| - \log|\alpha + \beta v|\right) = \frac{1}{\alpha}\log\left|\frac{v}{\alpha + \beta v}\right|.$$

Then, we get

$$v = \frac{\alpha}{\exp\left(\dfrac{\alpha}{m}t - \alpha C\right) - \beta}.$$

Employing the initial condition that $v = v_0$ when $t = 0$, we have

$$e^{-\alpha C} = \frac{\alpha + \beta v_0}{v_0}.$$

Substitution of this relation into the above solution for v yields

$$v = \frac{\alpha}{(\alpha + \beta v_0)\exp\left(\dfrac{\alpha}{m}t\right) - \beta v_0}v_0. \qquad \blacksquare$$

Example A.27. Under the influence of a uniform gravity along the negative z-axis, an ideal gas is enclosed in a cylindrical container at temperature T (Fig. A.14). Derive the pressure of the gas at height z, $p(z)$, where the molecular weight of the gas is μ and the gas constant is R. Let the acceleration of gravity be g and $p(0) = p_0$.

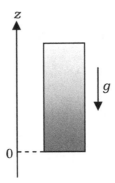

Fig. A.14.

Solution

Let a molarity (the number of moles in the unit volume) at height z be $\rho(z)$, the equation of state is

$$p(z) = \rho(z)RT. \qquad (A.18)$$

Let the base area of the container be A, and consider the balance of forces acting on the gas inside the infinitesimal region $A \cdot \Delta z$. The molarity $\rho(z)$ inside this infinitesimal region can be treated to be constant, so that we have

$$p(z)A = p(z + \Delta z)A + \mu \left[\rho(z)A\Delta z \right] g. \qquad (A.19)$$

Denoting the pressure change accompanied by the height difference Δz by $\Delta p = p(z+\Delta z)-p(z)$, we substitute Eq. (A.18) into Eq. (A.19) and employ replacements, $\Delta z \rightarrow dz$ and $\Delta p \rightarrow dp$, to obtain

$$\frac{\Delta p}{\Delta z} = -\mu\,\rho(z)g \quad \Rightarrow \quad \frac{dp}{dz} = -\frac{\mu\,g}{RT}p(z). \qquad (A.20)$$

Separating the variables (refer Sec A.7 Differential Equation 1) and using $p(0) = p_0$, we have

$$\int_{p_0}^{p} \frac{dp}{p} = -\int_{0}^{z} \frac{\mu\,g}{RT}dz \quad \Rightarrow \quad \log \frac{p}{p_0} = -\frac{\mu\,g}{RT}z,$$

from which we finally obtain

$$p(z) = p_0 \exp\left(-\frac{\mu g}{RT}z\right). \qquad \blacksquare$$

Index

Printed in the United States
By Bookmasters